Food Processing Operations and Scale-Up

FOOD SCIENCE AND TECHNOLOGY

A Series of Monographs, Textbooks, and Reference Books

1. Flavor Research: Principles and Techniques, *R. Teranishi, I. Hornstein, P. Issenberg, and E. L. Wick (out of print)*
2. Principles of Enzymology for the Food Sciences, *John R. Whitaker*
3. Low-Temperature Preservation of Foods and Living Matter, *Owen R. Fennema, William D. Powrie, and Elmer H. Marth*
4. Principles of Food Science
   Part I: Food Chemistry, *edited by Owen R. Fennema*
   Part II: Physical Methods of Food Preservation, *Marcus Karel, Owen R. Fennema, and Daryl B. Lund*
5. Food Emulsions, *edited by Stig Friberg*
6. Nutritional and Safety Aspects of Food Processing, *edited by Steven R. Tannenbaum*
7. Flavor Research: Recent Advances, *edited by R. Teranishi, Robert A. Flath, and Hiroshi Sugisawa*
8. Computer-Aided Techniques in Food Technology, *edited by Israel Saguy*
9. Handbook of Tropical Foods, *edited by Harvey T. Chan*
10. Antimicrobials in Foods, *edited by Alfred Larry Branen and P. Michael Davidson*
11. Food Constituents and Food Residues: Their Chromatographic Determination, *edited by James F. Lawrence*
12. Aspartame: Physiology and Biochemistry, *edited by Lewis D. Stegink and L. J. Filer, Jr.*
13. Handbook of Vitamins: Nutritional, Biochemical, and Clinical Aspects, *edited by Lawrence J. Machlin*

Other Volumes in Preparation

Food Processing Operations and Scale-Up

Kenneth J. Valentas
The Pillsbury Company
Minneapolis, Minnesota

Leon Levine
Leon Levine & Associates, Inc.
Minneapolis, Minnesota

J. Peter Clark
A. Epstein and Sons International, Inc.
Chicago, Illinois

MARCEL DEKKER, INC.   NEW YORK · BASEL · HONG KONG

Library of Congress Cataloging-in-Publication Data

Valentas, Kenneth J.
   Food processing operations and scale-up / Kenneth J. Valentas,
Leon Levine, J. Peter Clark.
       p. cm. -- (Food science and technology; 42)
   Includes bibliographical references and index.
   ISBN 0-8247-8279-8 (alk. paper)
   1. Food industry and trade. I. Levine, Leon, B.S. II. Clark, J.
P.   III. Title. IV. Series: Food science and technology
(Marcel Dekker, Inc.); 42.
TP370.V35   1990
664--dc20   90-49145
                                                         CIP

This book is printed on acid-free paper.

MARCEL DEKKER, INC.
270 Madison Avenue, New York, New York 10016

Current printing (last digit):
10 9 8 7 6 5 4

PRINTED IN THE UNITED STATES OF AMERICA

Preface

Food process engineering, which is a kind of process engineering, might be defined as what most chemical engineers and industrial or mechanical engineers are educated to do. That is, such engineers are educated to analyze, synthesize, design, and operate complex systems that manipulate mass, energy, and information to transform materials and energy into useful forms. In many chemical processes the materials of interest are gases or liquids subjected to chemical reactions, often modified by catalysts. Thus the typical process engineer is concerned with reaction rates and the thermodynamic properties of hydrocarbons.

If the useful forms in question happen to be human foods or animal feeds, the chances are good that they are solids, have a biological origin, and are transformed in processes that are not dominated by chemical reactions. Most engineers are not well prepared to solve the problems raised by such conditions. Food scientists, on the other hand, are well educated in the chemical, physical, and biological properties of foods, but generally have not been educated in the concepts of process engineering. It is common in industry for the food scientist and the engineer to work together on product and process development projects. Each is strong in some essential areas and weak in others. Fortunately, their strengths and weaknesses are different and so the team is balanced; however, communication can be difficult.

This book is intended to help students and practitioners who have a basic education in engineering (most likely chemical engineering) or food science to apply their skills to the challenges of the food industry. It contains some basic information in each area (engineering and food science) needed to permit communication, but it is not intended to substitute for more specialized courses or texts in either area.

It is based in part on a short course, "Food Processing Operations," taught for several years by Clark and Valentas for the American Institute of Chemical Engineers, and on a short course called "Process Scale-Up" offered by Levine.

This book has several purposes. The first is to describe some of the fundamental ideas of process development and design. The text discusses what questions to ask, why to ask them, how to answer appropriately, and how to deal with scale-up issues.

The second goal is to provide an engineering orientation to the food industry by presenting an introduction to the overall structure and how it works, consumer products, the chemistry of food ingredients and the economics of the food industry. We also want to discuss certain specific calculations and principles often neglected in other disciplines but important in food process engineering. Examples of these include solid material handling, evaporation, freezing, sterilization, cleaning and sanitation, fractionation by physical properties, extrusion, nutrition, sensory evaluation, and regulatory issues.

Very Imp.

To satisfy such disparate goals, we have modified the standard textbook approach by giving case studies and numerous examples of scale-up. Special care has been taken to present cases that are as real as possible. To avoid problems that could arise because of proprietary issues we have used only information that is in the public domain.

As industrial executives and consultants with academic and research experience, we bring an unusual point of view to a book that may be a classroom or self-study text. We are more interested in economy and efficacy than in precision and theoretical perfection. Appropriateness is the operative concept. Like the doctor's oath, "Do no harm," we say "Do what is necessary to support the decision at hand," "Ask the right question," rather than just doing what you know how to do.

Engineers in many fields sometimes do what is comfortable and familiar instead of what is helpful to business strategy. The engineer must learn to apply academic skills in the harsh commercial world of competitive business. The engineer's education begins in the classroom and continues in the practice of engineering in the world of commerce. This is a never-ending process.

This book compresses into one volume over sixty years of combined "on-the-job" learning as our careers have evolved in the dynamic food industry. If the reader grasps the principles that have served us so well and acquires some useful facts and techniques, we shall have accomplished our purpose.

Kenneth J. Valentas
Leon Levine
J. Peter Clark

Contents

1
Introduction

I. FOOD PROCESS DEVELOPMENT

What is process engineering? Process engineering is concerned with feasibility and practicality, that is, will something work and how much will it cost? The "something" in question is usually a process to manufacture a product (food product in this case) at a commercial scale. For any given product, there may be, and usually are, many conceivable methods of manufacture. Typical processes may be capable of manufacturing many different products. Establishing feasibility begins with the conception or selection of promising process candidates.

Since the food industry is oriented toward the retail consumer, in contrast to the oil or petrochemical industries, which have mostly industrial markets, there is a great emphasis on new product development, which generates a need for new processes or new applications of existing ones. This can mean that the first phases of food process development are directed at a purely imaginary product, for which there may not even be a prototype. Further, it may not be possible to have a prototype until a process is conceived.

More common than starting with an imaginary product, process engineering is often directed at scaling up from a bench-top (or kitchen stove) procedure, often first developed by a food scientist, to commercial scale. It is worth considering what commercial scale means in the food industry.

II. COMMERCIAL SCALE - ~~Size~~

Many food products are marketed in some form of convenient package or container, such as a carton, bottle, pouch, or other unit. These units range in size from a few ounces (in the case of confections) to several pounds (in the case of milk or soft drinks). A great many products are sold in units of about 1 pound (boxes of breakfast cereal, loaves of bread, cans of fruit and vegetables). Filling of discrete packages involves a number of separate steps, which are typically performed by mechanical equipment. A number of physical phenomena limit the currently achievable rates of such equipment. Among these limitations are friction, wear, acceleration of fluids, mechanical precision, flow rate of solids, and rates of heat transfer. Overcoming such limitations is one area of opportunity in food process engineering.

As a result of the typical rates achievable currently, commercially available filling equipment operates at rates ranging from 20 to 30 units/min up to 1200 units/min (in the case of high-speed can lines). For the typical 5-day, 2-shift operation of food plants, this gives theoretical production rates of 4.5–270 million units/year, for each filling line. That provides one measure of commercial scale.

Another measurement is financial. Total food sales (excluding restaurants and bulk agricultural commodities) in 1986 were $316.7 billion (3). The largest food companies have annual sales on the order of $10 billion. For these companies, a new product or family of products must have annual sales potential exceeding $100 million in order to represent an interesting opportunity. Not many food products have unit values exceeding $2, though average values are rising as emphasis is placed on greater convenience and higher quality. To achieve $100 million in annual sales, 67 million units (at $1.50 each) must be made. This would require one line operating at 300 units/min, three at 100 units/min, or six at 50 units/min. For high value products, the lower rates are more likely.

Commercial scale then, in the food industry, represents production rates of millions of units per year, scores of units per minute, and tens of millions of dollars in annual value from a typical production line. Multiple lines are common, as are multiple plants for nationally distributed products.

Feasibility is concerned with production of a given product at any rate (initially); scale-up is concerned with production at typical commercial rates when a slower and smaller volume process is known. Scale-up is discussed in some detail elsewhere (Chapter 10); it also is generally amenable to application of fundamental physical principles. Conception of original processes, on the other hand, is treated systematically only rarely. The text by Rudd, Powers and

Siirola (9) develops several heuristics, or rules of thumb, for inventing processes.

One purpose of a course or sequence in food process engineering, for which this volume might be used, is to provide the student or practitioner with a vocabulary and grammar for process synthesis in a new area. Extending the analogy to language, this suggests an emphasis on instruction in facts at first, followed by practice with their application. Some of the more relevant facts, and sources for obtaining additional ones, are discussed later. The balance of this book illustrates this approach with specific cases and chapters on certain essential fundamentals.

III. PRACTICALITY

Another important aspect of process engineering is practicality, that is, how much does a process cost to build and operate? In conventional chemical engineering, these questions are answered with well-established techniques developed over many years of experience with oil and petrochemical equipment and plants. The arsenal of process equipment types is relatively limited in many chemical plants (heat exchangers, pumps, towers, vessels, reactors, and some others), the principles of operation have come to be well understood and generalized through transport phenomena, and costs have been well correlated with simple measures of their size (8). Few of these conditions currently hold for food process engineering.

Although some food processing equipment is relatively simple (tanks and pumps, for example), much of it is highly specialized and complex. There is surprisingly little commonality among equipment developed for meat processing, baking, and dairies, though all are involved in large food processing segments. Another opportunity for progress in food process engineering would be generalization and simplification of equipment and processes across segments.

Progress has been made in understanding the fundamental principles of many important food process operations, but more remains to be done. The text by Loncin and Merson (6) is one of the best references on such fundamentals. A recent review by Schwartzberg (11) covers most of the available research literature on food process operations. The fact that such a review could even be written is one indication that the literature is relatively sparse.

Finally, food equipment tends to be fairly complex, for a number of reasons, and so correlations of cost with performance characteristics is difficult. It tends to be modular, and so there is little economy of scale, in contrast to much chemical equipment, where decreasing surface to volume ratios contribute to unit cost reductions as size is increased. When a chemical process is increased in rate, it is generally sufficient to make the tanks, reactors, and pumps larger. It requires less metal, all other factors remaining constant, to contain a unit volume as the volume increases. The major contribution to cost of most chemical process equipment is the amount of metal required.

By contrast, when a food process is increased in rate, it often is necessary to provide additional identical units, such as fillers. As a result, unit investment costs tend to be nearly constant, once commercial scale (as previously described) is achieved. In the face of these challenges, determination of food process costs proceeds along a path familiar to well-educated chemical engineers, but probably new to food scientists.

IV. FOOD PROCESS DESIGN

The approaches taught earliest in the engineering curriculum are among the most useful in food process design, especially heat and material balances. These relatively simple calculations to account for all material and energy flows are basic to process design. About 60–70% of the cost of most food products is the cost of raw materials, so yield improvement and proper accounting are critical to the efficient design and operation of food processes. Some yields can be surprising: It takes about 10 pounds of milk to make 1 pound of cheese; about 1-half the weight of a hog or steer is edible by humans; about one-half of the weight of an orange is juice.

Despite its importance, obvious to most engineers, preparation of an accurate material balance is not common in the food industry. Clearly, it should be. One complication is that many food processes are batch rather than continuous. Often batch and continuous operations are combined. This can lead to some conceptual and calculational challenges, because batch processing tends to be ignored in the typical engineering curriculum. At Purdue, Reklaitis, Okos, and their students (7,12,13) have done research on batch and semi-continuous process simulation, including some food examples.

Energy balances, likewise, are important, not because energy is a large cost in food processing, though it can be significant, but because correct application of energy may be the essence of a food process. Such important and unique processes as sterilization, pasteurization, freezing, drying, evaporation, baking, cooking, blanching, and frying all involve the addition or removal of heat to or from foods. Precise delivery of the correct amount of energy is important because both too much and too little can be harmful.

Because of the relative lack of known fundamental principles, physical properties, and rate constants, much food equipment sizing is done empirically, relying on experience or experiment. Rarely is experimentation as systematic as it could be, because the complexities appear overwhelming. It is important not to be overwhelmed in such cases, but rather to apply the statistical principles of experimental design and to focus on the critical parameters needed for process design, which tend to be related to rates of heat transfer, material flow, and biochemical reactions. There is an understandable tendency in much food process research to concentrate on the manufacture of edible product instead of useful data. This tendency arises from a confusion between the missions of process research and product development.

It has been said that the last thing one should do in studying baking is to bake bread. That does not mean never actually baking; rather, it means using model systems first to characterize heat transfer, learning the physical and chemical properties of dough, and developing mathematical models of heat and mass transfer in a system where geometry, weight, and properties all change with conditions and time. Then, bake bread to confirm the model.

In general, the development of a mathematical model is a good goal for food process development. Such a model provides a framework and a guide for research, serves as a useful tool for design once developed, and can demonstrate (by its relative success) the degree of understanding achieved. Relatively few food processes have been adequately modeled in this sense, certainly by comparison to other industries, and so such efforts represent another opportunity for progress in food process engineering.

V. ILLUSTRATIVE CASES

The food industry is very diverse, but it can be explored by sampling the major segments, as defined by the U.S. government in its Standard Industrial Classification (SIC). Table 1 lists the major segments in SIC 20, Food and Kindred Products. Four-digit SIC codes define more specific products within these broad segments. There are 47 four-digit codes currently in use. In contrast, there are 28 four-digit codes for SIC 28, Chemicals and Applied Products (14).

This volume includes relatively detailed descriptions of cases taken from several segments that are meant to provide a framework for further exploration in class or individual study. It is not meant to be a definitive reference for any process; such information sources abound in the patent literature and other texts. Information used

TABLE 1   Major Segments of SIC 20 Food *require*
different
process

201.   Meat and poultry products

202.   Dairy products

203.   Canning and preserving fruits and vegetables

204.   Grain mill products

205.   Bakery products

206.   Sugar and confections

207.   Fats and oils

208.   Beverages

209.   Miscellaneous and kindred products (seafood,
pasta, etc.).

here has been taken from such publicly available sources; it is not
necessary to use proprietary data to illustrate the approach. In
this introduction, a few additional cases are briefly discussed to
whet the appetite.

VI. POULTRY PROCESSING

Figure 1 is a block flow diagram prepared by A. Epstein and Sons,
Inc. (4) as part of an on-going study of energy consumption in the
food industry. Killing and dressing of poultry (chicken, turkey,
duck, game hens, etc.) is simpler than the corresponding steps for
red meats, such as beef and pork, but many of the issues are similar. Key characteristics of meat processing include the importance
of sanitation, the impact of regulation and inspection by government
agencies, opportunities for automation because of high labor intensity, and the significance of by-products to overall economics.
In the process, live animals are the raw material; animals that
die of other causes cannot be used and represent a yield loss. The
animals are immobilized by fastening to a conveyor chain, before killing in the case of birds and afterwards in the case of cattle and
hogs. By laws dictating humane slaughter, animals are stunned
electrically before they are killed by having their throats slit. (Exceptions are permitted to comply with kosher and Muslim religious
dietary codes; their ritual slaughter without stunning is defined as
humane.)

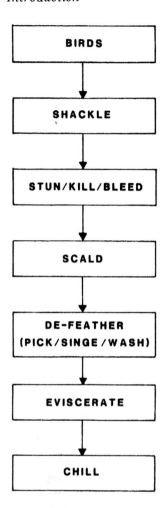

FIGURE 1   Poultry processing.

After blood is removed by draining (and collected for conversion
to food, adhesive, or feed), the remaining operations are designed
to remove inedible parts with minimal loss of meat. In the case of
birds, the focus is on feathers; for other animals, it is on hair and
hides. Scalding uses hot water to loosen feathers; which are then
beaten off in special equipment. Remaining pin feathers are singed
in a flame and the ashes washed off. (A similar process is used for

hogs to remove hair.)   The feathers are recovered for conversion
to animal feed.
   The entrails are removed manually, inspected to ensure absence
of disease, and separated into edible and inedible parts. The washed
and clean carcass is then chilled in ice for storage and shipment to
retail sale or further processing. Turkey and game hens are frequently wrapped in polymer film and frozen for sale. Chickens are
often cut into more convenient pieces for sale, fresh or frozen.
Whereas carcasses or large pieces of fresh meat were once the common commercial products, increasingly it is common to market more
highly processed, individual portions of partially prepared meat,
such as frozen, stuffed chicken breasts, deboned filets, or battered
nuggets. (The same trend extends throughout the food industry—
where cheese once was sold in large wheels, today it is sold in individual slices.)
   The animal processing industry is fiercely competitive, with low
profit margins, difficult labor relations, overcapacity, and relatively
old plants. Significant recent trends that characterize the industry
include the following:

A decline in per capita red meat consumption, probably due to perceived health concerns over hard fats found in beef and pork
Migration of the processing industry from traditional market centers
   (such as Chicago, St. Louis, and Kansas City) to areas closer to
   the Western feed lots (for cattle) or to the South (for chicken)
Increases in the per capita consumption of poultry and seafood
Increases in consumption of value-added meat products
Changes in ownership of many major meat processors as managers
   chose to focus either on meat packing (slaughter) or on processing

   Concerning this last point, Armour was sold by Greyhound to
Conagra, which then closed many slaughter plants; Swift was split
into two companies (Swift Independent Packing Company and Esmark,
which is now owned by Beatrice); General Foods (itself owned by
Phillip Morris) bought Oscar Mayer; and Iowa Beef Processors is
owned by Occidental Petroleum. Similar rearrangements have occurred in other segments of the food industry, but to a lesser extent. Where once meat packing companies were highly integrated,
in the sense that they slaughtered and processed, now most are
more specialized, performing one or the other function.
   Meat packing is one of the most highly regulated and inspected
of the food segments, in part because the original federal regulations were developed in response to an exposé of meat packing practices early in the 20th century (Upton Sinclair's The Jungle). The
U.S. Department of Agriculture controls meat processing, requiring

prior approval of plant designs, equipment, procedures, and chemicals. Slaughter and processing are conducted under continuous inspection by federal employees, usually educated as veterinarians. Because meat packing is labor intensive, unpleasant, repetitive, and occasionally dangerous, and humans can be a source of microbial contamination that contributes to spoilage, the industry is a good candidate for application of robots and automation. The negative trends cited earlier reduce the resources available for research and process improvement in the industry, but the opportunity is great.

VII. BEER

Figure 2 is a block flow diagram of the process for brewing beer, SIC 2082. It introduces some interesting biochemical engineering concepts common to many other industries, especially in connection with efforts to modify beer so as to reduce calories.

Beer is a beverage produced by fermenting certain cereal grains in water. It is an ancient art; there are paintings of Egyptians brewing and drinking a kind of beer made from barley. There are many varieties of beers and other malt beverages, differing in the strains of yeast used in the fermentation, the mixture of cereal grains, and the contributions of other ingredients.

Cereal grains are seeds; when they are moistened and maintained at a comfortable ambient temperature (70—80°F), they sprout or germinate in the first stages of becoming plants. The energy to form new cells is provided by converting starch stored in the grain to sugar. This conversion is catalyzed by enzymes that are activated by moisture and temperature and which are synthesized as the cells in the germinating seed reproduce. Properly controlled, this phenomenon can be used to convert the unfermentable starch, which is inaccessible to yeast, in cereal grains to fermentable sugars.

After a few days of germination, when the starch conversion enzymes have reached a high concentration, the process is halted by heating the moistened grain (kilning). Depending on the time and temperature used in kilning, the dried grain will have more or less color, resulting from the browning reaction of sugars and proteins and from carmelization. The various colors and flavors achieved at this step contribute to the unique character of subsequent beverages, such as light or dark beers, ale, porter, and stout.

Barley is the most common cereal used to make malt, but other grains can be malted, though they rarely are. Barley is a good source of enzymes but a relatively expensive source of starch. Thus, for economical and flavor reasons, cheaper sources of starch are combined with barley malt to make beer. The added materials are

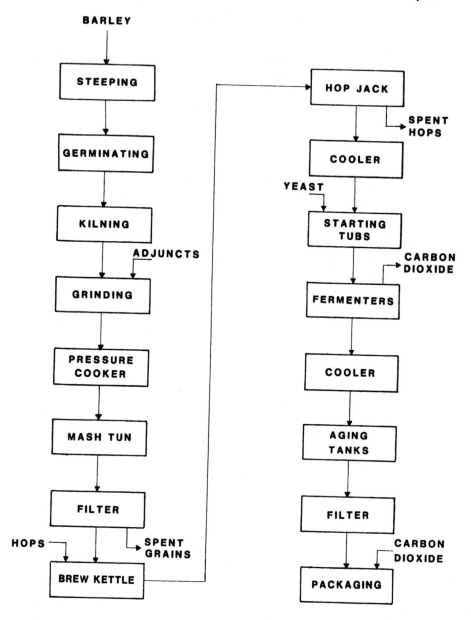

FIGURE 2 Beer production.

called adjuncts and may include corn, rice, wheat, cornstarch, dextrose, and corn syrup.

The mixture is ground to reduce particle size and cooked with water and steam under pressure to gelatinize starch and reactivate the conversion enzymes, which were temporarily halted when the sprouted grains were dried. The mixture, really a slurry that looks like hot breakfast cereal, is held in a vessel called a tun for several hours. The starch is converted by enzymes to sugar, which is dissolved, leaving behind insoluble components of the grains such as proteins, fiber, and fat. These are filtered out, and the solution is pumped to the brew kettle. The spent grains are dried and used as animal feed.

Hops are an aromatic herb grown in Europe and the northwestern United States. They contribute a characteristic bitter flavor to malt beverages, which is extracted by the hot clear mash (wort) in the brew kettle. Extracted hops are removed in the hop jack (a type of filter), and the wort is cooled and pumped to starting tubs, where yeast is added.

Breweries pride themselves on maintaining pure strains of distinct varieties of yeast, which contribute subtly to the flavor and may differ in yield and rate of fermentation. The yeast converts sugars to ethanol and carbon dioxide while also reproducing to generate more yeast cells. The fermentation produces heat because the basic chemical conversion is a type of oxidation, and so the fermenters are cooled to maintain a temperature between 40 and 58°F. A higher temperature produces off-flavors.

The carbon dioxide gas is usually compressed, dried, and sold, or used to carbonate the final product.

After fermentation, beer is aged for various lengths of time, usually several weeks. During aging of beer (and of other alcoholic beverages), a number of complex and poorly understood reactions occur that contribute to the final flavor. These include esterification, extraction, adsorption, and hydrolysis. Yeast cells settle out and many die, contributing some of their contents to the brew. Proteins from the yeast and the cereals may combine to form a haze or cloud, some of which settles out over time. If aging occurs in contact with wood, either in barrels or by addition of wood chips, there is some extraction of components, such as phenolics, from the wood by the brew and absorption of some components from the brew on the wood. (Whiskey and wine also undergo many of these mysterious reactions.)

After sufficient aging, the beer is filtered and packaged with addition of carbon dioxide to provide the characteristic effervescence. Typical packages are cans, bottles, and kegs. Packaging lines for beer are among the fastest in the food industry, reaching almost 2,000 cans per minute in some cases. Canned or bottled beer is usually heated after sealing to kill any residual yeast or bacteria that could cause disease or spoilage. This heat treatment, a type of pasteurization, which will be discussed in more detail later, has a detrimental effect on flavor, but it permits beer to be stored and shipped at room temperature. Beers intended for refrigerated storage and shipment, such as kegs of draft beer used in taverns, are not pasteurized and are considered superior in flavor to canned or bottled beer. Recently, the use of very fine filters (fractions of microns) to remove harmful microorganisms has resulted in packaged beers that do not require pasteurization to achieve room temperature storage stability.

VIII. LOWER CALORIE BEERS

Although beer has been a popular beverage literally for centuries, it has been more recently perceived as a contributor to excessive calories in the diet by some people, assisted in this perception by clever marketing to promote a new category of beers called "light." It is interesting to consider how light beers are made because the several techniques employed illuminate some of the variations possible in brewing.

There are many ways to reduce the calories in beer, and there is reason to believe that most have been used commercially (1). First, where are the calories in beer?

Alcohol contributes most, but not all of the calories in beer, from 94 to 103 in a 12-ounce can of typical domestic or imported beer, respectively out of a total of 160–198. In addition to the alcohol, typically 4–6%, residual unfermented carbohydrates are present. These are low molecular weight polysaccharides (dextrins) produced by the enzymatic hydrolysis of starch in the mashing step. The residual carbohydrates are digestible by humans but not by yeast, and they contribute to the viscosity (mouth feel) and flavor of the beer. Thus, the options for reducing calories in beer focus on reducing the alcohol, the residual carbohydrates, or both. The options that have been developed represent an interesting selection of process variations.

One popular approach is to add dextrose, which is completely fermentable, to the wort. This raises the concentration of soluble solids (measured as specific gravity) and lowers the residual carbohydrate content after fermentation. It also permits the production of more alcohol in a given volume of equipment and, since dextrose is cheaper than malt, actually reduces costs somewhat. The dextrose contributes little flavor of its own.

Certain enzymes can be used to reduce the dextrins to fermentable sugars by hydrolyzing branched polysaccharides. In Europe, the enzymes are added to the fermentation, and the fermentation time is extended. There is a risk of contaminating regular beers with the enzyme, which is heat stable and not destroyed by pasteurization. If this were to happen, the bottled beer would get sweeter in the bottle. An alternative is to add the enzyme before fermentation, then boil the wort to kill the enzyme, after it has had time to convert the carbohydrates to sugar.

During mashing, control of the temperature can manipulate the conversion of starch to sugar. The saccharifying enzyme, beta-amylase, is most active between 145 and 155°F, whereas the liquifying enzyme, alpha-amylase, is more active between 155 and 165°F. Holding the mash for up to 30 min at the lower temperature (145–155°F) increases the fermentable sugar concentration and gives higher alcohol and lower residual carbohydrate beers; shorter times at the lower temperature produces lower alcohol and higher residual solids beers. Near beer, 3.2% beer, and low-alcohol beers are produced this way.

The simplest approach to reducing calories is to add water, but unless the initial concentration of alcohol is higher than normal, the resulting product will not be acceptable. One way to achieve a higher initial concentration, permitting subsequent dilution, is to use a particularly active variety of malt (high diastase malt), which converts more of the starch in the malt and adjuncts to fermentable sugars, creating a higher concentration fermentation that can be diluted or reducing residual solids at normal concentrations. Both approaches are believed to be used commercially.

Finally, it has been proposed to use raw or refined sucrose as an adjunct since it is completely fermentable and may be cheaper than dextrose in some cases, though not in the United States. Raw sugar would contribute ash and color to the beer, which might be undesirable, whereas refined sugar would normally be expensive, but the technique would work.

In summary, beer, expecially reduced calorie or light beer, associated in the public mind with aggressive marketing and advertising, is the product of a surprisingly varied and sophisticated range of biochemical processes. The diversity teaches that there are often alternative routes to a desired product and the sophistication reminds us that even seemingly simple products can have significant science in their heritage.

IX. EDUCATION OF ENGINEERS FOR THE FOOD INDUSTRY

Food engineering is an academic program at a small number of universities, most often as a graduate program associated with departments of food science and agricultural engineering. The University of Massachusetts has the only Department of Food Engineering in the United States, a descendent of its agricultural engineering program, and at Colorado State University there is a Department of Agricultural and Chemical Engineering. There are food engineering options associated with departments of food science or agricultural engineering at the Universities of Illinois, California—Davis, Purdue, Cornell, Georgia, and a few others. Relatively few chemical engineering professors identify themselves as having research and teaching interests in food processing. Yet, chemical engineering is one of the better educational foundations for a career in food process engineering.

An education in food science can be the basis for work in food process engineering if properly supplemented with engineering courses. Since the emphasis is on engineering, it is probably more challenging to acquire this specialization by addition to a food science background than to do the reverse, but it can be done.

Ideally, a student with aspirations toward food process engineering would supplement his or her basic program in engineering (probably chemical or mechanical) with appropriate food science and biology electives. Obviously, the engineering curriculum designed to prepare graduates for success in electronics will not help much with food process engineering. However, foods and biochemistry do share many aspects, so electives taken in the general areas of biochemistry, nutrition, and microbiology would prepare an engineer to be uniquely useful in such industries as biotechnology and biomedicine, as well as in food processing.

One purpose of this volume is to help students and faculty gain an overview of the food processing industry from which they can pursue subjects of interest in greater depth, that is, to provide a context for further exploration.

The typical engineering curriculum is nearly filled with required courses and is stressed by the trend toward reduced required credit hours while technological advances introduce new candidates for study. Resolution of this dilemma is certainly not obvious, but if a student desires to orient his or her education toward the food industry, there are certain logical steps to take:

Take more hours than required to graduate, if possible, or pursue a graduate degree.

Use available science electives to take microbiology, biochemistry, and, where available, food science and nutrition. (There are about 35 departments of food science in the United States, mostly at land grant universities; there are about 135 departments of chemical engineering.)

Build a strong foundation in the core courses—heat and material balances, transport phenomena, unit operations, economics, and process control.

X. TRENDS IN MANUFACTURING

Beyond the educational fundamentals and industry-specific facts that, ideally, a food engineer in training should acquire, there are some general trends in industry that profoundly affect the food industry, but that are generally overlooked in universities. Numerous articles in the popular press have discussed the impact of Japanese management style on the automobile industry. Such concepts as Just in Time, statistical quality control, computer integrated manufacturing, and flexible manufacturing have entered everyday language, though not always with full understanding.

Forward-thinking food companies are applying the same concepts to their circumstances, but with some important differences. In some cases, food processing has been ahead of the times, for special reasons. For example, Goldratt and Cox (5) and Schonberger (10) view inventory as the enemy of efficient manufacturing, quite in contrast to conventional thinking, which focuses on long production runs and high equipment utilization. High inventory of work in progress and finished goods is the inevitable consequence. For many food products, high inventories are simply not possible because so many foods spoil so quickly. Thus, dairies, for many years, have made nearly every product every day, just as Goldratt and Cox (5) advocate. Process development in the dairy industry has been aimed at extending shelf life, even a small amount, to reduce the need for process changeovers, at the cost of higher inventory. Foods with longer shelf lives due to processing, such as canned or frozen, have been manufactured in fairly traditional fashion, with relatively long production runs and high inventories. Even in these industries, changes are occurring. Why?

Inventories create a number of problems:

First, they cost money—the value of the purchased materials used, which often amounts to 70% of the final value.

Second, bad things can happen to products held in process or in a warehouse. (In process, they are poorly protected, and after packaging, nearly all foods begin to decline in quality at some rate.)

Third, they reduce efficiency by creating queues in sequential manufacturing steps and so reducing response time to customer orders.

Fourth, they can reduce margins by requiring sales and promotions at reduced prices to move inventories that are in excess of market requirements (a very frequent occurrence!).

The great challenge in process design is to conceive processes that are highly flexible, have minimal work in progress, and produce increasingly sophisticated foods with zero defects, no waste, and at sufficient rates to respond to demand. Generically, such plants would be highly modular, fairly labor intensive (few devices are as flexible as a human), and very dependent on computer control.

Where automation is justified, as in material handling, where computer-controlled vehicles easily replace fork lifts, it will occur, and robots will replace humans in some food processing steps. However, it is likely that humans will always play an important role in food processing because so much of food quality is subjective, therefore uniquely dependent on human senses and judgment.

XI. SUMMARY

Food process engineering has been a rewarding engineering career for a number of people and offers great opportunities in the future as one of the emerging thrusts of chemical engineering. It is interesting to most people, because we all eat and can relate to the products, but it is also increasingly sophisticated as products and processes are developed to respond to society's demands for convenience, value, and good health. Faculty and students wishing to prepare themselves for careers in food processing should focus formal education on the fundamental biological sciences, which are usually neglected in engineering education, and undertake an informal education in the industry through visits, contacts, and outside reading. Worldwide trends in manufacturing are applicable to food processing in some unusual ways.

REFERENCES

1. Charm, S. E., *The Fundamentals of Food Engineering*, AVI, 1971.

2. Clark, J. P., Mathematical modeling in sterilization processes, *Food Technology*, 32:73 (1978).

3. *Food Engineering*, What gives the light to light beers? 50:101 (1978).

4. Gilson, P., *Industrial Market and Energy Management Guide, Food and Kindred Products*, American Consulting Engineers Council (1985).

5. Goldratt, D. M., and J. Cox, *The Goal: Excellence in Manufacturing*, Creative Output, Netherlands (1984).

6. Loncin, M., and R. L. Merson, *Food Engineering*, Academic Press, New York (1979).

7. Okos, M. R., and G. V. Reklaitis, Computer-aided design and operation of food processes in industry and academia, *Food Technology*, 39:107 (1985).

8. Peters, M., and K. Timmerhaus, *Plant Design and Economics for Chemical Engineers*, McGraw-Hill, New York, 1968.

9. Rudd, D. F., G. J. Powers, and J. J. Siirola, *Process Synthesis*, Prentice-Hall, Englewood Cliffs, New Jersey (1973).

10. Schonberger, R. J., *World Class Manufacturing*, The Free Press Division, Macmillan, New York (1986).

11. Schwartzberg, H., Process and plant design for food and biochemical production, in *The Frank Vilbrandt Memorial Volume for Chemical Process and Plant Design*, Y. A. Liu, ed., Wiley, New York (1986).

12. Shah, S. A., M. R. Okos, and G. V. Reklaitis, *A Computer Simulation Model for a Meat Processing Plant*, ASAE paper 83-6524 (1983).

13. Shah, S. A., M. R. Okos, and G. V. Rekalitis, *Production Scheduling in Food Processing Plants*, ASAE paper 83-6506 (1983).

14. Wei, J., T. W. F. Russell, and M. W. Swartzlander, *The Structure of the Chemical Processing Industries*, McGraw-Hill, New York (1979).

2
Overview of the Food Industry

Everyone has a vital interest in the food industry, and yet relatively few actually understand what it is, who the players are, and how it works. Humans have always been faced with the problem of feeding themselves to sustain life. Some cultures have been more successful than others, and, on average, we have learned how to manage nature's bounty. Basically, by trial and error, we have discovered how to stabilize a food supply so that many foods that are available in great abundance during a short period of time can be preserved and eaten between harvests. In effect, the food industry is the surge capacity in nature's cycle. With the fundamental problem of preservation resolved, issues such as wholesomeness, cleanliness, and safety could be addressed. When basic needs have been satisfied, products can be enhanced through improvement in quality, variety, or convenience.

Learning that wheat can be stored for long periods if properly dried solves the basic problem of preservation. Recognizing that water power can be used to turn a large millstone enhances the wheat by converting it to whole wheat flour. This adds convenience by taking the milling process out of the individual home and centralizing it. Developing a method for separating dark bran from the white endosperm further enhances the wheat and allows an entire array of new products to be developed so that variety is introduced. The baker adds convenience by centralizing the chore of daily baking and adds the ultimate in convenience by mechanically preslicing bread. In 1985, baked products in the United States recorded over $18 billion in sales (6).

Developments such as these have occurred over long periods of time. Basic food processing technology being utilized today had its

18

origin thousands of years in the past. Drying of foods is prehistoric. Processed meats were available in Europe in the Middle Ages. Milling and baking were known in 2600 B.C. Fermentation was practiced in 6000 B.C. (3). Canning, which is considered a contemporary food preservation technique, was discovered by Appert in 1809.

The engineer working in the food industry has spent a great deal of effort converting age-old processes to modern, automated manufacturing facilities that will deliver high quality, safe food products on a regular basis and at a reasonable price. This has not been an easy task, and it is truly remarkable that the industry has been so successful. As the food industry has developed to its present state, a highly organized infrastructure has resulted. This is shown schematically in Fig. 1 as the food pipeline, to draw an analogy from another industry.

Some basic products such as fish, fruits, and vegetables pass directly to the consumer with only minimal processing such as cleaning and packaging. These are handled by distributors and brokers who provide the interface between the farmers and supermarkets. The bulk of farm products are sold to "refiners." This group includes flour millers, meat packers, dairies, fish processors, oil refiners, and fruit and vegetable processors. A familiar line of commodities flows from the refiners through various distribution networks to the supermarket. These include flour, milled rice, milk, butter, cheese, cooking oils and fats, boxed blanched frozen vegetables, fresh meats, frozen raw fish, etc. The refiners also supply the value-added food processors with their ingredients. Many of the refiners are also value-added processors. Companies such as Armour, Swift, Land O'Lakes, Kraft, Pillsbury, General Mills, Procter and Gamble, and Staley (to mention just a few) supply themselves and others with ingredients to be converted to value-added products. Some examples of familiar products include baked goods, ready-to-eat cereal, cake mixes, salad dressings, soft drinks, frozen pizza, frozen entrees, snacks, fermented foods, beverages, and ice cream. The list of value-added products is long and continually changing. Although many of these products are household names, others are contemporary and rise and fall as lifestyles change and the economy fluctuates.

A very important position in the food pipeline is filled by food ingredient manufacturers. Their raw materials flow both from refiners of raw foods and from the fine chemical industry that specializes in food-grade functional chemicals such as artificial sweeteners, flow agents, antioxidants, gums, modified food starches, amino acids, and vitamins. The food ingredient manufacturers produce such important products as natural and artificial flavors and colors, seasonings, modified proteins and starches, specialty oils,

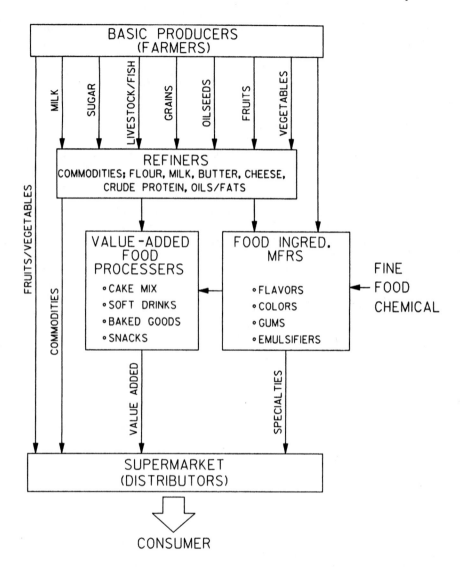

FIGURE 1   Food pipeline.

fats, and emulsifiers. Some of the companies involved in this segment of the business are also refiners and/or food processors in
the value-added segment. Others are strictly ingredient manufacturers and do not market a branded product to the consumer. Some

familiar names include Ralston Purina, American Maize, International Flavors and Fragrances, McCormick, Stauffer Chemical, and CPC. Products marketed directly through the supermarket would include items such as artificial sweeteners, spices and flavors, food colors, and dietetic foods.

Packaging is an integral part of today's food products. The food processing industry spent $34 billion for packaging materials in 1980. For many products (beer, soft drinks, breakfast cereals, frozen specialties, canned soups, baby foods, and pet foods), the packaging material can cost more than the ingredients (7). Packaging material suppliers work very closely with food manufacturers to develop a finished product that will maximize quality and protection of the food being delivered to the consumer.

The bulk of the foods sold in the supermarket have passed through the refiner/value-added route of the food pipeline. A relatively small number of companies have the awesome responsibility of feeding the nation safe, wholesome, tasty, and convenient foods without interruption. Think of the possible consequences if there were a food-related event similar to the oil embargo of 1974. One gets to the bottom line very quickly. Petroleum is important to our economy and life-style; food is absolutely essential to *life*.

I. SIZE OF THE INDUSTRY

The food industry is truly a giant. In 1985, it counted $307 billion in shipments (6), which does not include the fast food business with an additional $50 billion (11) in sales. For comparison, the U.S. automobile industry was $184 billion and the petroleum industry $436 billion (2). The food industry accounted for 14.5% of the gross national product in 1982 (3).

The major segments of the food industry are listed in Table 1. The $307 billion is divided into commodities and value-added products, with the split being about even. The commodity producers tend to be high-volume, price-competitive, productivity-driven industries with very little product differentiation. Value-added manufacturers strive for maximum product differentiation, are characterized by a marketing program that appeals directly to the consumer through advertising, and are continually improving existing products to stay in tune with the consumer's needs and desires.

II. FOOD INDUSTRY DYNAMICS

The food industry has been described as a mature, nongrowth industry. There is the occasional notoriety when a major acquisition

TABLE 1   Food Industry Shipments, 1985

Category	Value (billions $) Commodity	Value- Added	Comments
Meat	54.8	13.0	Primal cuts vs. frankfurters
Dairy	39.5	3.2	Fluid milk vs. ice cream
Fats/oils	12.7	5.6	Cooking oil vs. salad dressing
Animal feed	12.2	5.5	Livestock chow vs. soft moist dog food
Grain based	11.2	5.6	Family flour vs. ready-to-eat cereal
Frozen fruits and vegetables	6.0	7.2	Boxed blanched vegetables vs. boil-in-bag vegetables
Fish	4.9	2.4	Frozen 1-lb blocks vs. batter-coated pieces
Sugar/sweeteners	6.4		
Canned foods		15.0	
Baked products		20.7	
Soft drinks		19.7	
Coffee		5.9	
Candy		11.6	
Liquor		19.0	
Salad dressing/ condiments		5.8	
Pasta		1.2	
Dehydrated foods		1.5	
Misc. categories		16.4	
	$147.7	$159.3	
Total	$307		

Source: From Ref. 6, with permission.

occurs, but for the most part the food industry appears to be very stable and predictable. Yet, those who are part of it realize that it is dynamic and undergoing continual change. While per capita consumption is essentially constant on a mass basis of about 1,400 pounds per year, and volume growth mirrors population growth, the industry is quite dynamic within these boundaries (1). It is much like a system in the thermodynamic sense. Viewed from outside, an observer would conclude that nothing is happening. Viewed from within, there is continual movement and an endless struggle for market share and more of the fixed consumer dollar. New products are continually born and many old ones die as each of the companies strive to meet their individual growth objectives. The envelope of the food business is fixed in volume but elastic. Although some companies meet their aggressive growth objectives, others remain static or shrink. The desire for growth does not cause the dynamic situation that exists in the industry. It is driven by the consumer and their changing needs, tastes, and desires. Several factors are key.

III. CHANGING DEMOGRAPHICS

Several demographic factors have been impacting on food industry dynamics since the end of World War II. The United States has achieved near zero population growth. True tonnage growth has come to a virtual stop. The graying of America is upon us and will continue to exert more influence on food industry dynamics as senior citizens' needs are identified and satisfied by new products. The baby boomers have become sophisticated consumers with the purchasing power to support upscale products if the quality is high.

IV. CHANGING LIFE-STYLES

Many constants in our society are beginning to change. The number of working women is increasing markedly. Single-member households and households with two members working are increasing. Many meals are eaten on the run. "Grazing," that is, frequent snacking, is becoming a factor in our eating habits. Families are not eating together as frequently, and there is more demand for portion-size control. Leisure time is increasing. The microwave oven is now present in over half the households in the United States (8). Convenience is becoming even more important. Products such as Minute Rice; toaster waffles; frozen entrees; instant coffee and tea; boil-in-bag frozen vegetables; frozen orange juice; single-serve aseptically processed juices; and microwave pancakes, popcorn, and pizzas are designed to answer the demand for convenience. Superimposed on the demand for convenience is a growing desire for quality and a willingness to pay a premium price when the two needs are mutually satisfied.

V. HEALTH

The U.S. consumer is becoming more concerned over and aware of the relationship between food and health. Research at the medical/nutrition interface reveals new information almost daily. The information is disseminated to the public by the media almost immediately. A scientific article published in academic journals will often be syndicated by newspapers throughout the country. Unfortunately, the reporting is not always complete and objective, but the fact remains that public awareness of and interest in the food/health relationship is increasing. Products with the words "Natural" or "Lite" as part of their identity are viewed positively by consumers. The public is much more aware of the issues concerning preservatives, salt, cholesterol, and fat in the diet. Several products such as granola, yogurt, and margarine owe their success, at least in part, to the consumer's belief that they are healthy foods. The interest, discussion and controversy over artificial sweeteners in recent years is an indication of things to come in the future.

VI. GROWTH

Growth has become a byword in corporate board rooms throughout the United States, and the food industry is no exception. Security analysts have come to view any corporation that does not grow as stagnant and old-fashioned. Management measures its performance against the yardstick of growth in sales, profits, market share, return on invested capital, return on equity, and earnings per share.

If these are the ground rules, how can a company that is in a nongrowth industry grow? If the size of the pie is fixed, how can everyone increase their share? If you can't expand the envelope, the conclusion is obvious: Some will grow; others will shrink. The food industry has always been competitive, but the desire for growth increases the competitive pressure. The possible modes of growth can be classed in three broad areas: internal development leading to new products; acquisition and divestiture; and, finally, productivity and modern manufacturing (1).

Internal development leading to new products is not simple line extension. The objective is to develop major new volume with a product that does not exist today. Some recent examples come to mind: Pudding Pops, Cool Whip, and granola bars. One could argue that Pudding Pops is an elegant line extension of pudding, but it is more more than that. It requires no home preparation, is more like an ice cream novelty, and capitalizes on a "good for you" image. It is unique. It also required a considerable capital investment. Cool Whip is clearly a new product even though it mimics the "real" thing—whipped cream. It is technologically sophisticated and undoubtedly required a major research and development, and engineering investment to reach commercialization. Others have copied the product, but they are referred to as Cool Whip by the consumer.

Granola bars are not technologically sophisticated, and granola itself is not new. Many food companies have attempted to market granola products over the years, and many failed. The ultimate success of granola bars is actually a marketing success by General Mills. They selected the right positioning and the right name, and developed a product that was highly convenient, easily portable, good tasting, and perceived as highly nutritious. The name "Nature Valley Granola" is an elegant understatement of expert positioning. Since the success of General Mills with this product, others have introduced similar products to capture a slice of the pie, and market shares continue to shift in this argument.

VII. COMMITMENT TO RESEARCH AND DEVELOPMENT: ANNUAL REPORTS

New product development is expensive and risky. One can estimate the cost by noting the size of research and development expenditures for major food companies. Research and development as listed in annual reports range from $35 million to $130 million. Although it is certain that only a fraction of this is allocated to new products, even 25% would be a significant investment. In addition, the odds of success on new products, as opposed to line extensions, is low. Rules of thumb abound, and success figures such as 1 in 20 are not unusual for the industry as a whole. Still, the carrot is large, and even with high costs and low odds of success development of a new product is one of the driving forces for vitality in the industry.

VIII. ACQUISITION AND DIVESTITURE

Acquisition and divestiture has become quite common in the food industry. The magnitude of some more recent acquisitions and the resulting size of the "combined" companies is certainly significant. In 1981, the food industry recorded 666 mergers and acquisitions.

During the first 6 months of 1984, 327 had been reported (9). Beatrice paid $2.8 billion for Esmark, and the combined sales were $13.1 billion. Not to be outclassed, Nestle paid $3.0 billion for Carnation, with new combined sales of $14.8 billion. Prior to these megaacquisitions, Nestle acquired Stouffers to take a dominant position in the frozen entree business. Campbell's acquired Mrs. Pauls to become a factor in frozen fish. Pillsbury acquired Green Giant and, more recently, Haagen-Dazs to become the leader in premium ice cream.

The tobacco and alcoholic beverage industries have become major players in the acquisition game. Phillip Morris shocked the food industry when it acquired General Foods for a then record $5.75 billion to bring combined sales to $22.8 billion. R. J. Reynolds acquired Nabisco Brands, Inc. with combined sales of about $25 billion. Grand Metropolitan PLC, a British retailer of alcoholic beverages and pet foods, acquired The Pillsbury Co. for $5.75 billion. To add to its increasing influence in the value-added food business, Phillip Morris acquired Kraft for $13.5 billion to build a giant with combined sales of over $37.5 billion.

Why do companies acquire other companies? Is there a down side?

The advantages are as follows:

It is easier than building a business from scratch.
Immediate sales and profits appear in the annual report.
Synergies exist in distribution; there is consolidation of product
    lines; and leverage for shelf space in supermarkets occurs.
You can get rid of cash to avoid being a takeover target.

There is some downside risk in the acquisition route to corporate growth:

Acquisition prices usually exceed book value and often include a substantial amount for good will, trade names, etc. The result is a dilution of stockholder equity in acquiring company.
It is often disruptive to employees of both companies, especially when duplication must be eliminated.
It often requires excessive capital infusion especially in acquisition of smaller companies, which tend to be undercapitalized.
It often does not develop as expected.

There are two parts to this mode of growth. Along with acquisition is divestiture. This is often not given as much press unless there is a significant write-down of assets. But, it is a reality of the game.

IX. PRODUCTIVITY

Productivity is a straightforward strategy: Capture another's market share by outexecuting the competition in an existing business. The objective is to increase your productivity and improve margins to become the low-cost producer. The improved margins can be brought to the bottom line to meet profit objectives even without growth in volume.

How important is productivity? The chief executive of H. J. Heinz Co. challenged his managers to become low-cost competitors in the company's six largest businesses worldwide (10). CPC International increased its capital spending by 60% over a 3- to 5-year period to improve its worldwide manufacturing facilities by taking advantage of new technology to maintain low-cost, highest quality producer status (4).

The engineer plays a vital role in productivity improvement programs. There are three traditional areas of focus: labor reduction, yield improvement, and increased throughput. Endeavors in productivity improvement often involve significant capital investment. Heinz has announced plans to spend $110 million to automate a large plant in England. The objective is to double line speeds and reduce the work force by 1,200 people (11).

Automation implies labor reduction per unit of product. It does not necessarily mean robotics in the food industry. Certain technologies associated with the development of robotics are now and will continue to be applied to food manufacturing in the future. These include technologies such as computer vision. For robotics to be applied in the purest sense, two things must happen. Cycle times of robots must decrease by at least an order of magnitude. Current robot technology is too slow for food processing lines. Tactile sensors (touch, vision) must be improved to identify, judge, and manipulate food products, which are not uniform, as compared to machine parts or electronic components. Once these difficulties are overcome, the question then becomes the following: What is the cost?

Automation of the fixed variety continues to make an impact against labor reduction and throughput increase. Simulation tools now enable the engineer to conduct a bottleneck analysis on the computer so that improved machinery can be designed to increase throughputs. The impact of true automation goes far beyond faster machines and new machines to replace people. It also means true process control. This has not occurred in the food manufacturing industry to the same extent as it has in petroleum or chemical manufacture. Two things are lacking in food manufacturing: a thorough understanding of complex food processes that would permit dynamic modeling, and sensors and instrumentation that operate in real time to provide the correct signal for a process control loop.

In the past, computer capacity was the limiting factor that served also as an excuse to sidestep the issue of process control in food manufacturing. The bottleneck excuse has been eliminated by the silicon chip. The academic and industrial communities are beginning to come to grips with the two remaining challenges. It is not an insurmountable problem, but neither is it a trivial one. Applications will come sooner in fluid processing systems than in solid systems. Solids will be more difficult and will require innovative techniques in modeling and sensor development.

The benefits of good process control, whether it be closed loop or not, are many and include the following:

Reduction of downtime and off-grade product by elimination of process upsets and uncontrolled drift

Consistency of quality by elimination of process swings

X. MODERN MANUFACTURING

Food processing is just one player in the world of manufacturing. New concepts, generated by global competition in many markets, are emerging. These ideas represent major change in the philosophy of manufacturing and include concepts such as flexible manufacturing systems (FMS), computer integrated manufacturing (CIM), just in time (JIT), inventory control, and total quality control (TQC). The computer plays a vital role in this new strategy of competitive manufacturing and, in a sense, is the technology that makes it all possible.

Applications of these concepts in the food processing industry are only now beginning to be considered as manufacturer's become aware of the high cost of carrying inventory and the waste associated with inconsistent and uncontrolled processes. Processors are beginning to challenge traditional patterns that have, as their basis, large inventories of ingredients and finished goods. Increasing inventory turns not only reduces carrying costs but also the age of products presented to the consumer. For many products, quality deteriorates with time as well as temperature so that any reduction in product age will likely have a positive effect on product quality.

REFERENCES

1. Behnke, J. R., Food Technology, August, p. 24 (1983).
2. Business Week, March 22, p. 56 (1985).

3. Connor, J. M., R. T. Rogers, B. W. Marion, and W. F. Mueller, *The Food Manufacturing Industries*, Lexington Books, Lexington, Massachusetts (1985).
4. Cook, J., *Forbes*, May 7, p. 59 (1984).
5. *Food Engineering*, November, p. 12 (1985).
6. *Food Processing*, February, p. 22 (1986).
7. Gallo, A., and J. M. Connor. Packaging in food marketing, *National Food Review*, 17:10−13 (1981).
8. *Microwave World*, December, p. 5 (1985).
9. Business event: mergers, *Prepared Foods*, November, p. 80 (1984).
10. *Restaurant Business*, September 20, p. 188 (1986).
11. Saporito, W., *Fortune*, June 25, p. 44 (1985).

3
Financial Analysis of Capital Projects

The primary purpose of investing capital in plants and equipment is to earn an amount of money that exceeds the investment. The importance or value of such an investment is measured by the "return." Return can be expressed in many ways, the most common being:

Average annual rate of return
Payback period
Discounted cash flow return on investment (ROI)
Net present value (NPV)

In any business environment, projects and ideas must compete for limited capital dollars. Decisions must always be made regarding the relative merit of proposals for capital expenditures to sustain or insure profits. Projects that are dictated by compliance with laws and regulations, good manufacturing practice, product safety, employee welfare, disasters or accidents, and the like, are generally not evaluated on the basis of a financial return. However, projects intended to add to profitability must be measured in an objective manner.

There are generally two questions to resolve:

1. Given a capital proposal, does it meet minimum financial objectives of the company?
2. Given several alternative capital projects, which offers the best "return"?

To discuss financial analysis in an abstract manner is laborious and difficult. The concept of "return" is best introduced by way of example.

Example 1

A project has been proposed to invest $100,000 for new equipment that will generate a yearly net cost savings of $30,000 before taxes. The equipment has an economic life of 5 years, and the company requires a minimum ROI of 10% on cost savings projects. The income tax rate is 50%.

Does this project meet the company's minimum financial hurdle of 10% ROI? What is the actual ROI?

How is ROI calculated? What about the other measures of "return" such as average annual rate of return and payback? How does one handle taxes and depreciation?

First consider average annual rate of return. On a pre-tax basis the computation is

$$\text{Average annual return} = \left[\left(\frac{150,000 - 100,000}{100,000}\right)/5\right]100\% = 10\%$$

What about payback period? On a pre-tax basis it is

$$\text{Payback} = \frac{100,000}{30,000} = 3.3 \text{ years}$$

This is a simple problem in that the yearly profit is constant over the life of the project. Yet both of these methods ignore a very critical aspect of financial analysis, namely the time value of cash flow.

To illustrate the concept of time value, consider the investment of $100 in a savings account that pays 10%, compounded annually, for a period of 5 years. The cash flow over the 5-year period would be as follows:

Year	Cash
0	$100 (invest)
1	100 + 10 = $110
2	110 + 11 = 121
3	121 + 12.1 = 133.1
4	133.1 + 13.3 = 146.4
5	$146.4 + 14.6 = $161.0

The formula for compound interest will give this result directly,

$$S = P(1 + i)^n \tag{1}$$

where

S = future sum
P = present value
i = annual interest rate
n = number of years

Then, $S = \$100(1 + 0.1)^5 = 100(1.61) = \$161$.
This function is given in tabular form (Table 1) for various values of i and n.

TABLE 1   Future Values of $1. $S = P(1 + i)^n$

Periods	2%	4%	6%	8%	10%
1	1.0200	1.0400	1.0600	1.0800	1.1000
2	1.0404	1.0816	1.1236	1.1664	1.2100
3	1.0612	1.1249	1.1910	1.2597	1.3310
4	1.0824	1.1699	1.2625	1.3605	1.4641
5	1.1041	1.2167	1.3382	1.4693	1.6105
6	1.1262	1.2653	1.4185	1.5869	1.7716
7	1.1487	1.3159	1.5036	1.7138	1.9488
8	1.1717	1.3686	1.5938	1.8509	2.1436
9	1.1951	1.4233	1.6895	1.9990	2.3589
10	1.2190	1.4802	1.7908	2.1589	2.5938
11	1.2434	1.5395	1.8983	2.3316	2.8532
12	1.2682	1.6010	2.0122	2.5182	3.1385
13	1.2936	1.6651	2.1329	2.7196	3.4524
14	1.3195	1.7317	2.2609	2.9372	3.7976
15	1.3459	1.8009	2.3966	3.1722	4.1774
16	1.3728	1.8730	2.5404	3.4259	4.5951
17	1.4002	1.9479	2.6928	3.7000	5.0545
18	1.4282	2.0258	2.8543	3.9960	5.5600
19	1.4568	2.1068	3.0256	4.3157	6.1160
20	1.4859	2.1911	3.2071	4.6610	6.7276

In evaluating an investment opportunity, attention is focused on the present value of a future receipt. The concept of present value and discounted cash flow is illustrated by rewriting Eq. 1 in the form

$$P = \frac{S}{(1 + i)^n} \qquad (2)$$

Let's apply this concept to a typical situation. You are trying to sell a speedboat for $15,000. A potential buyer offers to pay you $8,000 now and an additional $9,000 in two years. If you could earn 10% interest on money deposited in a bank, should you accept the offer? You really need to know the present value of the offer. How much is the future payment worth in today's dollars? By applying Eq. 2,

$$\text{Present value} = \$8,000 + \frac{\$9,000}{(1 + 0.1)^2}$$

$$= \$8,000 + \$9,000(0.826) = \$15,434$$

The offer exceeds the asking price by $434. This represents a premium for accepting the risk that the buyer might default on payment.

I. ROI (RETURN ON INVESTMENT)

The concept of discounting is easily extended to annual payments, which is usually the case in a capital project proposal.

By investing an amount of money, $X now, you can receive a cash payment of $1 at the end of each year for 3 successive years. If you desire to earn 10% on your investment, what is the value of $X? (The choice of a specific discount rate is a complex issue, which is discussed later. For now use 10%.)

The cash flow from each year must be discounted back to time zero so its value can be determined in the present.

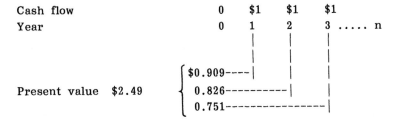

Cash flow	0	$1	$1	$1	
Year	0	1	2	3	..... n

Each dollar is discounted by the factor $[1/(1 + i)]^n$ for the appropriate year. In this example, $i = 0.1$, or 10% per year.
The amount $X is given by the present value calculation as $2.49.
Reversing the logic, if an investment of $2.49 generates a cash flow of $1/year over 3 years, the ROI is 10%.
The concept can be generalized to define ROI:

$$P_j = \frac{S_j}{(1 + i)^j} \tag{3}$$

where

P_j = present value from the jth year
S_j = cash flow in jth year
i = discount rate or ROI

Figure 1 illustrates the flow of funds in a typical business operation on which a return might be calculated. The various terms in this diagram will be explained in more detail later.
Then, ROI is that value of i for which

$$\sum_{j=1}^{n} P_j = \text{investment} \tag{4}$$

and n = number of years. The factor $[1/(1 + i)]^j$ is called the "present value factor." These are listed in Table 2 and plotted in Fig. 2 for typical values of i.

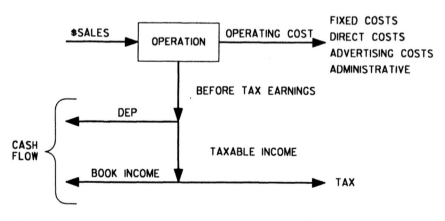

FIGURE 1   Cash flow diagram.

Calculation of ROI is by trial and error to satisfy Eq. 4. Many calculators are programmed to perform this calculation. Now, the original problem can be attacked to compute the actual ROI. In this case,

Investment = \$100,000
n = 5 years
Depreciation = straight line for 5 years
Income tax rate = 50%

The solution is illustrated next with the definition of some terms that recur in financial analysis (Fig. 1):

Taxable income = before tax earnings −depreciation
TI = BTE −D
Cash flow = before tax earnings −taxes
CF = BTE −T
Book income = taxable income −taxes
BI = TI −T

The difference between book income and cash flow is the depreciation. Depreciation is an allowance permitted by the Internal Revenue Service to cover replacement cost of a fixed investment. It helps to reduce income taxes and represents a contribution to cash flow. Then:

Year	BTE	D	TI	T	CF	BI
1	30,000	20,000	10,000	5,000	25,000	5,000
2	>	>	>	>	>	>
3	>	>	>	>	>	>
4	>	>	>	>	>	>
5	30,000	20,000	10,000	5,000	25,000	5,000

ROI can now be estimated by finding that value of i for which Eq. 4 is satisfied. In this case, values of i of 10% and 8% were selected for trial. The results are shown in tabular form. Present value factors were obtained from Table 2.

	$i = 10\%$			$i = 8\%$		
Year	$\left(\dfrac{i}{1+i}\right)^j$	CF	$\dfrac{S_j}{(1+i)^j}$	$\left(\dfrac{i}{1+i}\right)^j$	CF	$\dfrac{S_j}{(1+i)^j}$
1	0.909	25,000	22725	0.926	25,000	23150
2	0.826	25,000	20650	0.857	25,000	21425
3	0.751	25,000	18775	0.794	25,000	19725
4	0.683	25,000	17050	0.735	25,000	18375
		S = 94725			S = 99700	

TABLE 2 Present Value Factors. $1/(1 + i)^n$

n	2%	4%	6%	8%	10%	12%	14%	16%
1	0.98	0.962	0.943	0.926	0.909	0.893	0.877	0.862
2	0.961	0.925	0.89	0.857	0.826	0.797	0.769	0.743
3	0.942	0.889	0.84	0.794	0.751	0.712	0.675	0.641
4	0.924	0.855	0.792	0.735	0.683	0.636	0.592	0.552
5	0.906	0.822	0.747	0.681	0.621	0.567	0.519	0.476
6	0.888	0.79	0.705	0.63	0.564	0.507	0.456	0.41
7	0.871	0.76	0.665	0.583	0.513	0.452	0.4	0.354
8	0.853	0.731	0.627	0.54	0.467	0.404	0.351	0.305
9	0.837	0.703	0.592	0.5	0.424	0.361	0.308	0.263
10	0.82	0.676	0.558	0.463	0.386	0.322	0.27	0.227
11	0.804	0.65	0.527	0.429	0.35	0.287	0.237	0.195
12	0.788	0.625	0.497	0.397	0.319	0.257	0.208	0.168
13	0.773	0.601	0.469	0.368	0.29	0.229	0.182	0.145
14	0.758	0.577	0.442	0.34	0.263	0.205	0.16	0.125
15	0.743	0.555	0.417	0.315	0.239	0.183	0.14	0.108
16	0.728	0.534	0.394	0.292	0.218	0.163	0.123	0.093
17	0.714	0.513	0.371	0.27	0.198	0.146	0.108	0.08
18	0.7	0.494	0.35	0.25	0.18	0.13	0.095	0.069
19	0.686	0.475	0.331	0.232	0.164	0.116	0.083	0.06
20	0.673	0.456	0.312	0.215	0.149	0.104	0.073	0.051

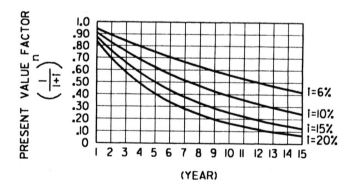

FIGURE 2 Present value factor.

18%	20%	22%	24%	26%	28%	30%	35%	40%
0.847	0.833	0.82	0.806	0.794	0.781	0.769	0.741	0.714
0.718	0.694	0.672	0.65	0.63	0.61	0.592	0.549	0.51
0.609	0.579	0.551	0.524	0.5	0.477	0.455	0.406	0.364
0.516	0.482	0.451	0.423	0.397	0.373	0.35	0.301	0.26
0.437	0.402	0.37	0.341	0.315	0.291	0.269	0.223	0.186
0.37	0.335	0.303	0.275	0.25	0.227	0.207	0.165	0.133
0.314	0.279	0.249	0.222	0.198	0.178	0.159	0.122	0.095
0.266	0.233	0.204	0.179	0.157	0.139	0.123	0.091	0.068
0.225	0.194	0.167	0.144	0.125	0.108	0.094	0.067	0.048
0.191	0.162	0.137	0.116	0.099	0.085	0.073	0.05	0.035
0.162	0.135	0.112	0.094	0.079	0.066	0.056	0.037	0.025
0.137	0.112	0.092	0.076	0.062	0.052	0.043	0.027	0.018
0.116	0.093	0.075	0.061	0.05	0.04	0.033	0.02	0.013
0.099	0.078	0.062	0.049	0.039	0.032	0.025	0.015	0.009
0.084	0.065	0.051	0.04	0.031	0.025	0.02	0.011	0.006
0.071	0.054	0.042	0.032	0.025	0.019	0.015	0.008	0.005
0.06	0.045	0.034	0.026	0.02	0.015	0.012	0.006	0.003
0.051	0.038	0.028	0.021	0.016	0.012	0.009	0.005	0.002
0.043	0.031	0.023	0.017	0.012	0.009	0.007	0.003	0.002
0.037	0.026	0.019	0.014	0.01	0.007	0.005	0.002	0.001

The ROI is about 8%, which is below company standards of 10%. This project would not be acceptable. This example is particularly straightforward because the cash flow is constant for each year, which greatly simplifies the calculation. Recalling from Eq. 3,

$$\sum_{j=1}^{n} P_j = \sum_{j=1}^{n} \frac{S_j}{(1 + i)^j}$$

If S_j is constant,

$$\sum_{j=1}^{n} P_j = s \sum_{j=1}^{n} \left(\frac{1}{1 + i}\right)^j \tag{5}$$

Table 3 gives values of $\sum_{j=1}^{n}[1/(1 + i)]^j$ as a function of i and n. This is very useful in problems where S is constant or can be assumed constant as a first approximation. For example, from Table 3, the value of $\sum_{j=1}^{n}[1/(1 + i)]^j$ for i = 0.08, and n = 5 is 3.993, which gives $\Sigma P_j \simeq 100,000$.

II. TIMING AND ROI

The importance of timing is clearly illustrated in Table 4, which shows several cash flow scenarios and the resultant ROIs.

Note that each scenario would have the same average annual pretax return: 14%. However, the pre-tax ROIs vary from a low of 16% to a high of 26%. Discounting places a high value on cash flow in the early years of the project and a low value in later years as can be seen by comparing cases I and II. Variability in cash flow will also reduce ROI depending on how severe the valleys and peaks become. Case IV is particularly interesting in that it occurs with regularity on new, large-scale projects. It is not uncommon on such projects to be late on start-up due to factors such as construction delays, equipment delivery problems, and inability to meet a predetermined start-up curve. This can have a devastating effect on ROI as can be seen by comparing cases IV and I. Case IV is identical in cash flow to case I but with a 1-year offset.

The concepts that have been introduced here can best be brought together through an example that is typical of the food processing industry.

Example 2

The new venture department wishes to conduct a feasibility study on a food product recently developed by research and development, and successfully consumer tested by marketing research. As the project engineer, you must determine the economic viability of the proposal. A preliminary engineering study indicates the need for a phased capital investment with $1,700,000 initially and an additional $500,000 in the beginning of the third year of the project to meet the capacity projected for the sales predicted by marketing.

These assets have a 10-year economic life and are to be depreciated on a straight line basis.

Sales projection

Year	1	2	3	4	5	6	7	8	9	10
Sales ($1,000)	750	1,500	4,000	5,000	5,600	6,000	6,400	6,700	7,000	7,200

TABLE 3   Cumulative Present Value Table. $\sum\limits_{j=1}^{n}\left(\frac{1}{1+i}\right)^j$

n	8%	10%	12%	14%	16%	18%	20%	22%	24%	25%	26%	28%	30%
1	0.93	0.91	0.89	0.88	0.86	0.85	0.83	0.82	0.81	0.80	0.79	0.78	0.77
2	1.78	1.74	1.69	1.65	1.61	1.57	1.53	1.49	1.46	1.44	1.42	1.39	1.36
3	2.58	2.49	2.40	2.32	2.25	2.17	2.11	2.04	1.98	1.95	1.92	1.87	1.82
4	3.31	3.17	3.04	2.91	2.80	2.69	2.59	2.49	2.40	2.36	2.32	2.24	2.17
5	3.99	3.79	3.60	3.43	3.27	3.13	2.99	2.86	2.75	2.69	2.64	2.53	2.44
6	4.62	4.36	4.11	3.89	3.68	3.50	3.33	3.17	3.02	2.95	2.88	2.76	2.64
7	5.21	4.87	4.56	4.29	4.04	3.81	3.60	3.42	3.24	3.16	3.08	2.94	2.80
8	5.75	5.33	4.97	4.64	4.34	4.08	3.84	3.62	3.42	3.33	3.24	3.08	2.92
9	6.25	5.76	5.33	4.95	4.61	4.30	4.03	3.79	3.57	3.46	3.37	3.18	3.02
10	6.71	6.14	5.65	5.22	4.83	4.49	4.19	3.92	3.68	3.57	3.46	3.27	3.09
11	7.14	6.50	5.94	5.45	5.03	4.66	4.33	4.04	3.78	3.66	3.54	3.34	3.15
12	7.54	6.81	6.19	5.66	5.20	4.79	4.44	4.13	3.85	3.73	3.61	3.39	3.19
13	7.90	7.10	6.42	5.84	5.34	4.91	4.53	4.20	3.91	3.78	3.66	3.43	3.22
14	8.24	7.37	6.63	6.00	5.47	5.01	4.61	4.26	3.96	3.82	3.69	3.46	3.25
15	8.56	7.61	6.81	6.14	5.58	5.09	4.68	4.32	4.00	3.86	3.73	3.48	3.27
16	8.85	7.82	6.97	6.27	5.67	5.16	4.73	4.36	4.03	3.89	3.75	3.50	3.28
17	9.12	8.02	7.12	6.37	5.75	5.22	4.77	4.39	4.06	3.91	3.77	3.52	3.29
18	9.37	8.20	7.25	6.47	5.82	5.27	4.81	4.42	4.08	3.93	3.79	3.53	3.30
19	9.60	8.36	7.37	6.55	5.88	5.32	4.84	4.44	4.10	3.94	3.80	3.54	3.31
20	9.82	8.51	7.47	6.62	5.93	5.35	4.87	4.46	4.11	3.95	3.81	3.55	3.32

TABLE 4   Cash Flow Dynamics

Investment		I. Decreasing cash flow $50,000	II. Increasing cash flow $50,000	III. Variable cash flow $50,000	IV. 1-Year delay decreasing cash flow $50,000
	1	19,000	5,500	10,200	0
	2	16,500	8,000	8,900	19,000
	3	15,000	8,900	16,000	16,500
	4	13,600	10,200	7,100	15,000
	5	12,000	11,700	14,500	13,600
	6	10,600	12,500	7,700	12,000
	7	9,800	13,400	12,000	10,600
	8	8,900	15,700	14,500	9,800
	9	7,950	17,200	13,000	8,900
	10	6,650	16,900	16,100	7,950
	11				6,650
Total cash flow		120,000	120,000	120,000	120,000
Net profit		70,000	70,000	70,000	70,000
Average annual pretax profit		7,000	7,000	7,000	7,000
Average annual return		14%	14%	14%	14%
Payback years		3	5.5	4.5	4
ROI		26%	16%	18.7%	19.4%

Total production costs for products of this type are 45% of sales; administrative and sales costs are 25% of sales. You anticipate a start-up expense of $250,000 the first year plus research and development and engineering expenses of $100,000 in year 1 and $50,000 in year 2. The income tax rate is 50%.

What is the ROI?

Placing the data in tabular form is helpful in keeping things straight and also consistent with most spread-sheet software, such as Lotus 1-2-3*, and other packages for personal computers. This is shown in the cash flow table (Table 5).

TABLE 5 Cash Flow Table ($1,000)

	Year									
	1	2	3	4	5	6	7	8	9	10
Capital	1,700		500							
Income	750	1,500	4,000	5,000	5,600	6,000	6,400	6,700	7,000	7,200
Expense Production, administration, and sales	525	1,050	2,800	3,500	3,920	4,200	4,480	4,690	4,900	5,040
Start-up	250									
Research and development, and engineering	100	50								
ΣExpenses	875	1,100	2,800	3,500	3,920	4,200	4,480	4,690	4,900	5,040
BTE	-125	400	1,200	1,500	1,680	1,800	1,920	2,010	2,100	2,160
Depreciation	170	170	220	220	220	220	220	220	220	220
TI	-295	230	980	1,280	1,460	1,580	1,700	1,790	1,880	1,940
Tax	-147.5	115	490	640	730	790	850	895	940	970
Cash flow	22.5	285	710	860	950	1,010	1,070	1,115	1,160	1,190

The trial and error calculation for ROI (i) is also shown in tabular form (Table 6), utilizing present value entries from Table 2 with assumed values for i of 24% and 28%.

One loose end must be reconciled. The second infusion of capital at the end of year 2 is not fully depreciated at the end of the project. That portion of the capital has 2 years of life remaining on the books after year 10. This represents a positive cash flow of $100,000 in year 10 and must be discounted by the appropriate factor from year 10 and added to the CDCF. Thus for i = 24%,

CDCF = 2383 + (0.116)(100) = 2,395

and for i = 28%,

CDCF = 1928 + (0.085)(100) = 1,937

Recalling the criterion for determining ROI,

ΣP_j = CDCF = investment

But the same logic that was applied to the undepreciated assets in year 10 must be applied to the infusion of $500,000 capital after the first 2 years of the project. The $500,000 must be discounted by the appropriate factor from year 2.
Then for

i = 24%; investment = 1,700 + 0.65 (500) = 2,025

i = 28%; investment = 1,700 + 0.61 (500) = 2,005

Two iterations on i are generally sufficient to estimate ROI by linear interpolation. In this case, 27% is close enough.

It is very easy to become overly precise with discounted cash flow calculations. Considering the accuracy of the data, a high degree of precision is hardly warranted. The ROI from such a calculation should only be viewed as an order of magnitude estimate. No amount of sophisticated financial analysis can improve "bad" numbers. Sensitivity analysis should always be performed to test the strength of the proposal. It really amounts to playing the "what if" game. What if:

The capital estimate is low by 30%?
Start-up is difficult, and sales are reduced by 50% in the first
   year?
The steep increase in the sales curve between years 2 and 3 does
   not occur as planned?

TABLE 6 Calculation of ROI

	Year									
	1	2	3	4	5	6	7	8	9	10
Cash flow ($1,000)	22.5	285	710	860	950	1,010	1,070	1,115	1,160	1,190
i = 0.24										
PVF[a]	0.805	0.650	0.524	0.423	0.341	0.275	0.222	0.179	0.144	0.116
DCF[b]	18	185	372	364	324	278	237	200	167	138
CDCF[c]	18	203	575	939	1,263	1,541	1,878	2,078	2,245	2,383
i = 0.28										
PVF[a]	0.781	0.610	0.477	0.373	0.291	0.227	0.178	0.139	0.108	0.085
DCF[b]	18	174	339	321	276	229	190	155	125	101
CDCF[c]	18	192	531	852	1,128	1,357	1,547	1,702	1,827	1,928

[a]PVF = present value factor.
[b]DCF = discounted cash flow.
[c]CDCF = cumulative discounted cash flow.

These and other issues can be quantified relative to the original
ROI estimate of 27%.
For feasibility studies, capital estimates are not usually closer
than ±30%. The impact of a 30% increase in capital can be estimated by referring back to Table 5. Items above the line will not
be changed. The only item affected below the line is depreciation.
The new capital is ($1.7 × 10^6)(1.3) = $2.21 × 10^6 at inception, and
(1.3)($0.50 × 10^6) = $650,000 after the second year. The revised
figures are shown in Table 7.
By trial and error, it can be shown that the ROI for this scenario is 22%. This is a notable decrease from the original estimate
of 27%, and indicates that the proposal may be sensitive to capital
and the quality of the capital estimate will be an issue of concern.
Many cost savings or productivity proposals can be quickly evaluated through ROI analysis. Consider the following example, which
occurs frequently in the manufacture of frozen products.

Example 3

Fresh shrimp are being flash frozen in a fluidized bed cryogenic
freezer utilizing liquid CO_2 as the heat transfer medium. Shrimp
enter the freezer at 50°F, have a freezing point of 26°F, and exit
the freezer at 10°F. The CO_2 freezer and storage tanks are currently being leased for $2,000/month. The plant has an hourly pro-
duction rate of 10,000 lbs/hr and operates 2,000 hr/year. Liquid CO_2
will provide 100 BTU of cooling per pound of CO_2 and costs $0.03/lb.
It has been proposed to invest $1,200,000 for a mechanical re-
frigeration fluidized bed freezer to replace the cryogenic freezer.
The mechanical freezer has a heat loss factor (infiltration, fans, etc.)
of 39 tons of refrigeration (1 ton refrigeration = 12,000 BTU/hr).

TABLE 7   Cash Flow ($1,000) for 130% Capital

					Year					
	1	2	3	4	5	6	7	8	9	10
BTE	-124	400	1,200	1,500	1,680	1,800	1,920	2,010	2,100	2,160
DEP	221	221	286	286	286	286	286	286	286	286
TI	-346	179	914	1,214	1,394	1,514	1,634	1,724	1,814	1,874
Tax	-173	89	457	607	697	757	817	862	907	937
Cash flow	+48	310	743	893	983	1,043	1,103	1,148	1,193	1,223

The refrigeration compressors require 5 HP to produce 1 ton of refrigeration. Electricity costs $0.06/kw-hr. The maintenance cost of the mechanical refrigeration is 4%/year of the capital investment. Assuming 10-year straight line depreciation and a 50% income tax rate, is this proposal worth pursuing?

The physical data are as follows: λ = heat of fusion (water) = 144 BTU/lb; Cp_f = heat capacity shrimp (frozen) = 0.5 BTU/lb-°F Cp_u = heat capacity shrimp (unfrozen) = 0.65 BTU/lb-°F, water content of shrimp = 78%.

The heat load is the same with either refrigeration system: heat load = Q = sensible heat + latent heat.

$$Q = Cp_u (T_{in} - T_{freeze}) + Cp_f (T_{freeze} - T_{out}) + \lambda x_w$$

where x_w = mass function of water in the shrimp.
From the data and constants given,

$$Q = 0.65(50-26) + 0.5(26-10) + 144 (0.78) = 136 \text{BTU/lb}$$

and on an hourly basis, $Q = [(136)(10,000)]/12,000 = 113.3$ tons of refrigeration.

A side-by-side comparison of cost factors is helpful in arriving at the difference in operating cost between the two alternatives:

	Mechanical refrigeration	Cryogenic refrigeration
Capital	$1,200,000	$0
Energy	Load = 113.3T + 39T = 152.3T Electric = (5 HP)(0.746)(152.3) = 568 kw	1 lb CO_2 = 100 BTU cooling CO_2 = (1.36 × 10⁶) (2,000)/100 = 2.72 × 10⁷ lb/year
Cost	(568)(0.06)(2,000) = $76,092/year	2.72 × 10⁷ ($0.03) = $816,000/year
Other	Maintenance @ 4% of capital = $4,800	Tank lease = $24,000/year
Total operating cost	$124,092/year	$840,000/year

The yearly savings with the mechanical system is then $723,840. In reality, this savings will vary from year to year due to inflation and other factors. However, for the purpose of obtaining a first-order estimate of ROI it is reasonable to assume constant savings over the life of the project. Then, Eq. 5 applies, with

P = $12,000,000
n = 10 years

Savings = $723,840
DEP   = $120,000/year
TI   = $723,840 - $120,000
Tax   = $301,920
CF = S = $421,920

Substituting in Eq. 5,

$$1,200,000 = 421,920 \sum_{j=1}^{10} \left(\frac{1}{1 + i}\right)^j$$

and

$$\sum_{j=1}^{10} \left(\frac{1}{1 + i}\right)^j = 2.84$$

From Table 3, reading across the row for n = 10 and interpolating, the ROI is approximately 33%.

This appears to be an attractive proposal. However, there are some issues related to the difference in freezing rate between CO_2 and mechanical air blast processes. Freezing is rapid with CO_2 as compared to air blast. Users of CO_2 claim two advantages as a result of this. First, the rapid freezing with cryogenic mediums causes ice crystals in the product to be small. Slow freezing produces larger crystals with more damage to cellular structure and some difference in textural quality. This is somewhat subjective and difficult to quantify in economic terms. And if product is subjected to any freeze-thaw cycles in the distribution system it may not be an important issue. The second issue is quantifiable and is related to the fact that slower freezing in air blast systems results in greater weight loss due to evaporation of moisture from the product during freezing. With a cryogenic system, this might only be 0.5%, but is normally 3% in blast freezing and claimed to be as high as 8% with fish (2). This could have a significant effect on the economics of the proposal.

Assuming the shrimp to have a value of $1.50/lb and an incremental moisture loss of 1-1/2%, this would amount to 300,000 lb/year or $450,000/year. This changes the economics considerably. The yearly savings are now reduced to $265,908. Working through Eq. 5, the calculated ROI would be only 10% from Table 3.

The economic merit of the proposal is highly sensitive to the evaporative moisture loss. Extensive and accurate tests would have to be performed to determine the actual loss. This in itself might prove difficult unless a scalable size blast freezer were available for extensive testing.

There are other factors that would enter into the final decision. More detailed comparisons of cryogenic and mechanical freezing systems can be found in the literature (1). An option worth considering is a combination of cryogenic and mechanical freezing. Cryogenics can be utilized to encrust the product with an ice layer that will greatly reduce evaporative moisture loss. Freezing is then completed with a smaller mechanical system.

The intent here is to show that a simple ROI calculation is a useful tool in comparing alternatives.

III. MINIMUM ROI

The minimum ROI, or hurdle rate, is not a universal constant but in fact varies by industry and by company within a given industry. It is a number that each corporation selects individually. In general terms,

ROI = f(pure rate, risk factors)

where the pure rate is the rate of interest just needed to induce investment in absence of considerations such as loss risk and inflationary risk.

Risk factors include

Inflation
Business risk
Competitive pressure
Political risk/monetary policy
Change in tax laws

Purposely, no numbers have been indicated for these factors. Neither are these factors all-inclusive; they are only an indication that hurdle rates "evolve" in a given corporation, and once that occurs they become the benchmark by which all projects are measured.

In any event, most corporations will demand a return on investment that substantially exceeds the return available from safe investments and the cost of borrowing funds at the most favorable rates, often called the "prime rate."

IV. NPV (NET PRESENT VALUE)

ROI analysis is particularly useful for determining if a specific project meets the company's hurdle rate but not always sufficient

for selecting from among projects that all exceed the hurdle rate.
Suppose, for example, that the hurdle rate is 15% and you are trying to differentiate between two projects with ROIs of 15% and 25%,
respectively. Which would you choose? The project with the highest ROI may not be the right choice.

ROI measures only one dimension of a financial proposal. Since
it is actually a ratio, it ignores the absolute magnitude of the opportunity. For example, how would you select from two projects
that had the same ROI, say 25%? One way around this apparent
dilemma is to measure the cash flow generated at the required
hurdle rate. This introduces the concept of NPV:

$$NPV = present\ value\ -\ investment \tag{6}$$

or, in terms of previous definitions,

$$NPV = CDCF\ (@\ hurdle\ rate)\ -\ investment \tag{7}$$

NPV is a measure of the amount of cash generated in excess of
the investment plus interest computed at the hurdle rate. Consider
the projects listed in Table 8.

It would appear that the automatic case packer has the greatest
NPV. In the absence of any risk, this would be the appropriate
choice. However, one cannot ignore risk. Installation of the automatic case packer, if it is either new technology or simply new only
to your company, will have some risk associated with it. Compara-
tively, a new warehouse to eliminate charges for rented space
should have virtually no risk. The difference in NPV is $300,000.
How much risk are you willing to take for $300,000?

This relatively simple example illustrates the importance of considering both risk and NPV in choosing between competitive projects that otherwise exceed corporate hurdle rates. As much work

TABLE 8   Projects @15% Hurdle Rate

	Investment	CDCF (@15%)	NPV
Automatic case packer (labor reduction)	$1.2 MM[a]	$1.8 MM	$0.6 MM
New warehouse (elimi- nate outside charges)	$2.0 MM	$2.3 MM	$0.3 MM

[a]MM = million.

as is practical should be done to quantify risk levels. In the specific example, experiments on the equipment in question and/or contact with other companies using the same equipment in similar applications would be appropriate.

V. BUSINESS PERSPECTIVE

ROI as it has been defined and illustrated here is a very useful tool for evaluating what is often referred to as a "tactical" business opportunity. Accounting tools and, in particular, ROI have a tendency to be tactical rather than strategic.

As has been illustrated in the example problems, ROI analysis has a tendency to place more importance on near-term profits rather than long-term reward. This is a mathematical reality of discounted cash flow analysis, and is fine for deciding on the relative merits of competing proposals or the financial merit of a new product, as these are tactical decisions.

However, many strategic decisions that a business might consider have poor cash flow streams in the early years but eventually lead to growth in future years. Consider for a moment whether some of the Japanese companies that eventually became dominant in cameras, electronics, or automobiles would have committed to such long-term projects on the basis of ROI analysis. Many years of investment spending, with poor cash flow, were required to nourish these strategic business decisions. One cannot argue with their ultimate success.

ROI analysis can be a powerful tool if used judiciously.

REFERENCES

1. Briley, G. C., *ASHRAE Journal*, Dec., pp. 30—32 (1980).
2. Eckett, A., *Food Processing*, June, p. 28 (1985).

4
Food Chemistry

I. SCOPE OF CHAPTER

Those engaged in processing food should understand the basic chemistry and physics of food materials. Often the engineer in the food processing industry has acquired just enough knowledge of food chemistry to size a heat exchanger, design a piping system, and so on. This information has often been obtained in an unstructured way dependent on need and is essentially a random walk in the world of food chemistry. This chapter is designed to pull together some basic concepts of food chemistry that are of particular interest for engineering applications. For those desiring a more indepth discussion the reader is referred to Fennema's (6) textbook, which is detailed and also contains a wealth of useful references.

II. ENZYMES AND HUMAN NUTRITION

It is difficult to discuss food in an entirely objective manner because everyone has likes, dislikes, prejudices, and opinions. For those of us engaged in processing food, it is important to understand the basic chemistry and physics of food materials. The ultimate purpose of food is to sustain biological systems, and, in this case, the human organism is the prime concern. There is considerable controversy over the role of foods and food ingredients in human nutrition. Nutrition is a very complicated science and the subject of extensive research. It is not possible to treat the subject in any depth here. Still, it is important for engineers to understand the fundamentals of human nutrition even if on a very elementary level.

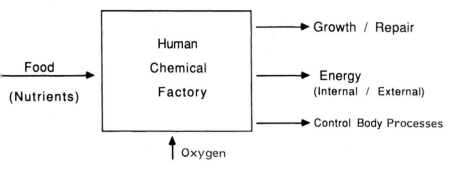

FIGURE 1   Human chemical factory.

III. HUMAN CHEMICAL FACTORY

It is instructive to view the human body as a chemical factory with specific inputs and outputs, such as shown in Fig. 1. In this simplified model, foods or nutrients that can be categorized chemically as lipids, proteins, carbohydrates, water, minerals, and vitamins along with oxygen, which is obtained from the air we breathe, are utilized by the body to support life. In essence, these foods permit the human chemical factory to convert food to the compounds necessary to (a) provide for growth and repair of the organism, (b) supply the body with energy for external and internal process, and (c) control the many complex processes occurring in the body.

The analogy to a chemical factory can be carried one step further to focus on the manner in which foods are utilized (Fig. 2). Foods such as carbohydrates and proteins are large macromolecules that must be reduced to smaller units to be useful in the metabolic processes. The body is not unlike a refinery where the first step is to break down large complex molecules to simpler forms and then recombine and reform them into compounds that can be utilized for maintenance of the body.

As indicated in Fig. 2, carbohydrates are large macromolecules with molecular weights on the order of 10,000–5,000,000. Proteins that are consumed as food have molecular weights on the order of 60,000. In the process of digestion, the body reduces these large carbohydrate and protein molecules to simple sugars and amino acids with molecular weights on the order of 100–150. All this happens in a period of about 20 hr. How does this remarkable phenomenon occur?

A. Enzymes

Enzymatic hydrolysis is the breakdown mechanism for macromolecules in the digestive process:

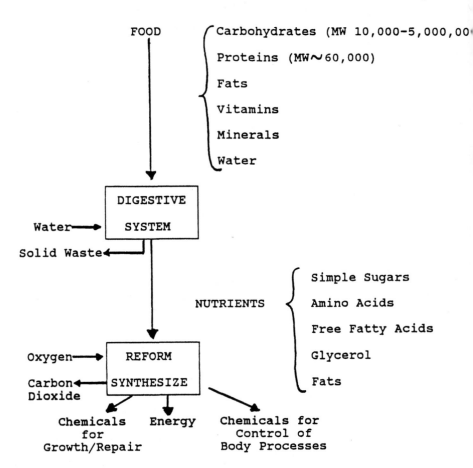

FIGURE 2   Breakdown and synthesis processes in human chemical factory.

$$AB + H_2O \xrightarrow{\text{Catalyst}} AOH + BH$$

The catalyst is an enzyme that is a complex protein capable of accelerating chemical reactions by factors of $10^{12}-10^{20}$ over uncatalyzed reactions at low temperatures ($25-50°C$).

The exceptionally high efficiency of enzymes at low temperatures is important not only in the digestive process but also in many food processing operations.   For example, enzymes can cause flavor, color odor, and texture changes in many foods over a wide moisture range and even at subfreezing temperatures.

Enzymes are named after the substrate on which they act. Thus, the enzyme glucose isomerase will act on glucose to convert it to the isomer, fructose.

GLUCOSE ISOMERASE

GLUCOSE   FRUCTOSE

This is a very important reaction commercially. Fructose is significantly sweeter than glucose. Glucose, which is derived from cornstarch by the enzymatic action of α-amylase and glucoamylase, can be enzymatically converted to the sweeter isomer fructose. A mixture of approximately 45% glucose and 55% fructose has a sweetness comparable to sucrose and is less expensive than sucrose. It is called "high fructose corn syrup" (HFCS) and is used extensively in soft drinks as a sucrose replacement to reduce cost. The economic impact on the soft drink industry is quite significant.

From a food processing standpoint, some of the more important types of enzymes are as follows:

Hydrolases
   starch $\xrightarrow{\text{amylases}}$ maltose
   maltose $\xrightarrow{\text{maltase}}$ glucose
   fats $\xrightarrow{\text{lipases}}$ fatty acid and glycerol
   proteins $\xrightarrow{\text{peptidases}}$ amino acids

These reactions are essential for human nutrition.

Oxidases
   phenolase (browning)
   lipoxygenase (off-flavors)
   ascorbic acid oxidase

Oxidases are mentioned because the reactions which they catalyze are generally undesirable.

Phenolase is responsible for enzymatic browning in fruits and vegetables, which occurs when phenols are oxidized to orthoquinones, which then polymerize to form brown compounds. The methods that have been developed to reduce or destroy phenolase activity other than blanching involve the use of sulfur dioxide or sulfites. Acidulants such as citric acid can be used to control browning as long as the pH is maintained at 3 or lower. The age-old remedy of squeezing lemon juice on fresh cut fruit to prevent browning depends on reduction of pH to inactivate phenolase.

Lipoxygenase adds oxygen to polyunsaturated fatty acids, esters, and glycerides, resulting in undesirable off-flavors. The enzyme is generally inactivated by blanching.

Ascorbic acid oxidase, which is naturally present in citrus juices, will oxidize ascorbic acid (vitamin C). Thus, pasteurization prior to packaging and freezing is important to minimize the loss of this important nutrient.

In addition to naturally occurring enzymes, food processors add manufactured enzymes to foods for various reasons. Some of the more common applications are as follows:

1. Pectinases

These enzymes (containing polygalacturonase, and pectin methylesterase) are all used in the clarification of fruit juices and wines.

2. Invertase

This enzyme hydrolyzes sucrose to glucose and fructose, and is widely used in confectionary manufacture. It prevents grittiness and crystallization and is particularly useful for manufacture of coated candies with semifluid centers such as chocolate-covered cherries.

3. Proteases

These enzymes (containing bromelain, ficin, or papain) are useful for hydrolysis of proteins for applications such as meat tenderization, manufacture of protein hydrolysates, and fish protein processing.

4. Rennets

These are used to coagulate milk in cheese manufacture.

5. Amylases

These generally hydrolyze starch and are useful in baking, brewing, and manufacture of sweetener from starch.

B. Specificity/Selectivity

The high degree of selectivity of enzymes is essential to living cells and, from a less esoteric viewpoint, of practical importance in food

processing. Most enzymes will attack only one substrate and catalyze only a single reaction. The selectivity of enzymes is often enhanced by substances called coenzymes. Coenzymes are often vitamins or closely related to vitamins. The B vitamins, for example,
are important in the biochemical process of manufacturing the coenzymes required by the body. The combination of an enzyme and
coenzyme is referred to as a conjugated protein, that is, the combination of a protein molecule and a nonprotein molecule or prosthetic group. For example, phenolase enzymes need copper as a
prosthetic group to catalyze browning reaction in fruits and vegetables.

C. Temperature Effects

Enzymes are unique because they function in the biologically important temperature range of 25—45°C. In this range, the temperature
relationship is essentially Arrhenius. At temperatures above 45°C,
enzymes begin to become inactivated due to denaturation of the constituent protein. Thermal inactivation occurs over a temperature
range dependent on the specific enzyme. Thus, in a mixture of
enzymes, one enzyme might be totally inactivated at 50°C, whereas
another might not become inactivated until the temperature reached
60°C. Activation energies for catalysts are on the order of 6,000—
15,000 cal/mole, whereas activation energies for denaturation or enzyme inactivation are on the order of 50,000—150,000 cal/mol (6).

   If one is concerned with enzyme inactivation, as is most often the
case in food processing, the rate of heat transfer and the temperature of the heat transfer medium are important issues. In particulate systems (fruits, vegetables), the geometry of the pieces as well
as mode of heat transfer must be considered in process design. Undesirable components can be enzymatically produced if the rate of
temperature rise from ambient to inactivation temperature is too slow.
Conversely, enzyme inactivation may not be uniform throughout the
piece if conduction is significantly rate limiting.

   At subfreezing temperatures, most enzymes are not inactivated
and many show measurable activity in systems that are considered
frozen. This is of particular concern with regard to shelf life of
some frozen foods. Enzymes such as lipoxygenase and phenolase
could cause undesirable changes in fruit and vegetables if not inactivated by blanching prior to freezing.

   Blanching of corn on the cob is an essential step prior to freezing. The difficulty of blanching corn on the cob is due to the size
and geometry of corn. In abstract terms, the process is transient
heat transfer from a bulk phase fluid (hot water or steam) to a cylinder. The cylinder is a composite material consisting of a cob with
an outer layer of kernels, which undoubtedly will have different

physical properties. This problem has been modeled (16) using peroxidase activity as an indication of adequacy of heat treatment. To simplify the mathematics, the corn cob was considered to be homogeneous, which is a reasonable assumption under these circumstances. In addition, heat transfer resistance at the surface was considered to be negligible (BIOT >100, or heat transfer coefficient >300 BTU/hr ft² °F), which simplifies the boundary conditions and permits solution of the unsteady state heat conduction equation in closed form. The validity of this assumption is dependent on the extent of agitation in the bulk phase. In spite of this simplifying assumption, the mathematics are still cumbersome and require iterative procedures for numerical solution. The model presented also takes into account the heat inactivation that occurs during the cooling period after heating. Table 1 shows the form of the solution with average percent peroxidase activity as a function of heating/cooling time for blanching temperatures of 80°C and 90°C.

The significance of these data is the large difference in enzyme activity between the kernels and the center of the cob. Enzyme activity in the cob material can lead to undesirable odor and flavor development upon extended storage of the frozen product. In a case such as this, geometry is a critical variable in heat conduction and should not be ignored. The obvious solution is to overblanch to be certain of inactivating all enzymes, but this is not necessarily optimum. Excessive blanching can result in quality loss due to overheating even though enzymes are inactivated. It has been suggested that peroxidase should not be used to measure adequacy of blanching since it is more heat resistant than suspected spoilage enzymes (17). In green peas, lipoxygenase is

TABLE 1   Peroxidase Activity (Percent) as a Function of Heating/Cooling Time

	Time (min)	Kernel	Center of cob
80°C	8	75	95
	16	50	90
	24	30	80
90°C	8	60	95
	16	20	80
	24	10	70

Source: From Ref. 16, used with permission.

shown to be a more appropriate indicator of the adequacy of heat treatment if one is interested in appropriate enzymes deactivation with minimum heat treatment (9,28).

The freeze-thaw cycles that occur in any distribution system for frozen foods contribute negatively to this problem by exposing portions of the product to elevated temperatures for short periods, thus allowing enzymatic reaction to proceed at higher rates during these periods. Time/temperature indicators could be useful for monitoring quality changes due to enzymatic and nonenzymatic reactions. Their efficacy has been demonstrated for monitoring quality in frozen strawberries (21). With appropriate quality models and statistical sampling techniques, these devices could be useful in inventory management programs in distribution centers.

D. pH

Many enzymes are most active in the pH range of 4.5–8.0. There are notable exceptions such as pepsin, which prefers an acid medium, and trypsin, which prefers alkaline medium. Both enzymes hydrolyze protein.

Undesirable enzymatic reactions, such as the action of phenolase on fruits, can be controlled by utilizing pH.

E. Digestion/Hydrolysis

The human digestive system is essentially isolated from the body in the sense that food which is ingested orally is broken down by hydrolysis or emulsified as in the case of fats and finally absorbed by the bloodstream through the intestinal walls. The digestive process itself can be simplistically viewed as a series of chemical reactions as depicted in Fig. 3.

The process begins with the sight, aroma, and taste of the food. These organoleptic signals cause the salivary gland to excrete mucin, which acts as a lubricant, amylase to begin the hydrolysis of carbohydrates, and salt to activate the amylase. Chewing serves the purpose of size reduction and mixing of food and saliva. The mixture is transported down the esophagus by peristaltic action and enters the upper portion of the stomach where it comes into contact with gastric juices excreted through the stomach lining. The gastric juices contain pepsin, which hydrolyses protein to smaller peptones, rennin, which coagulates milk, and hydrochloric acid, which is necessary for pH control. As with saliva, the flow of gastric juices is stimulated by organoleptic factors of aroma, sight, and taste. It can also be stimulated chemically by meat extractives commonly found in meat-based soups. Psychological factors such as anxiety, fear, and depression can inhibit the flow of gastric juices. Between the

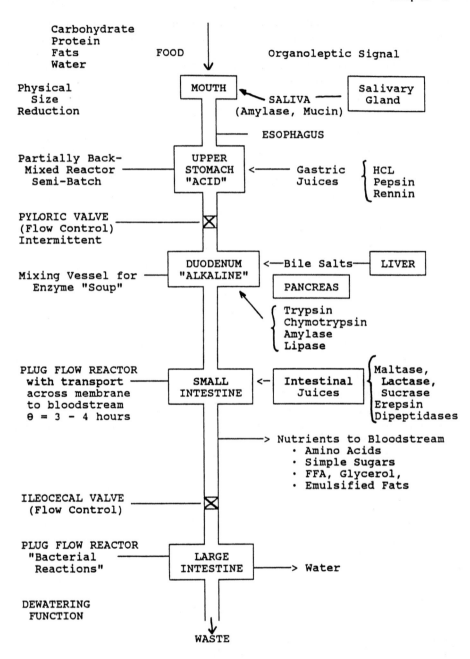

FIGURE 3   Flow sheet of digestive system.

upper stomach and the duodenum, the pyloric valve, which serves as a check valve to separate the two vessels, opens intermittently to transfer the reaction mixture to the duodenum. In the duodenum, the mixture changes from acidic to alkaline. Bile salts, which aid in the emulsification of fats, are introduced from the liver. Pancreatic juice, which contains trypsin and chymotrypsin to complete the hydrolysis of proteins to peptones, amylase, to convert polysaccharides to maltose, and lipase, for hydrolysis of some of the fat, are also added. Finally, intestinal juices are introduced through the intestinal wall downstream of the duodenum. The intestinal juice contains maltase, sucrase, and lactase to reduce disaccharides to monosaccharides and erepsin to hydrolyze peptones to dipeptides and dipeptidase to complete the hydrolysis of amino acids. Flow through the small intestine is essentially plug flow and is caused by muscular action. During transport through the small intestine, sugars, amino acids, and hydrolyzed and emulsified fats are transferred through the intestinal walls to the bloodstream. These nutrients then become the raw materials for the metabolic processes. The residence time from ingestion to the end of the small intestine is 3 to 4 hr. After passing through the small intestine, undigested food, unabsorbed nutrients, and waste material passes through the ileocecal valve into the large intestine. The large intestine's primary function is the removal of water from this mass. The large intestine is also rich in bacteria. These may act on undigested nutrients such as oligosaccharide or lactose with the subsequent production of gas and resultant flatulence. Fiber, which is consumed as a natural part of many foods, functions much like a filter aid due to its bulking action. The residence time in the large intestine is on the order of 20 hr.

The power of enzymes is remarkable in that such large macromolecules as repesented by carbohydrates and proteins are rapidly reduced to simple sugars and amino acids in a short time.

IV. PROTEIN

Proteins are essential for proper human nutrition. The body has a specific need for fixed nitrogen, which it cannot satisfy from elemental or inorganic sources. Proteins are large, complex macromolecules. For instance, a naturally occurring material such as casein is a mixture of many different proteins. From a nutritional standpoint, the body requires certain amino acids to be present in varying amounts in all protein sources. Synthesis of protein in the body as well as other nitrogen-containing substances required to sustain life can be accomplished provided the necessary amino acids are contained in sufficient quantity in protein consumed as food.

Before discussing the structure and properties of proteins, we will provide an overview of how nitrogen, in the form of protein, enters into the life cycle. This is depicted in Fig. 4. Nitrogen

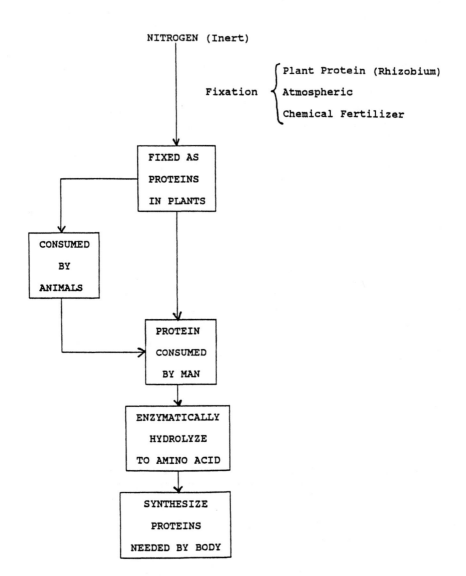

FIGURE 4   Nitrogen source for human nutrition.

is plentiful in the atmosphere, but for human nutritional purposes it is inert. Nitrogen is chemically fixed atmospherically by electrical discharge (lightning) with the subsequent formation of nitrates, which find their way to the soil during rainfall. It is also fixed in chemical plants in the form of chemical fertilizers, which are mechanically applied to the soil. The third, and perhaps most elegant, is direct fixation by leguminous plants such as soybean, peas, beans, clover, and alfalfa. This occurs because of a symbiotic relationship with the soil bacteria *Rhizobium japonicum*, which fixes nitrogen in the root nodules. Although poorly understood, the potential of bacterial nitrogen fixation, if it could be applied to nonlegumes such as wheat, would have dramatic worldwide benefits.

Once nitrogen is fixed as a protein in plants, it enters the human food chain either through direct consumption of the plants or their seeds or indirectly through the consumption of animals that have eaten the plant/seed protein. When the protein is consumed by humans, it is hydrolyzed in the digestive process, enzymatically, to the constituent amino acids. In the metabolic processes, the body utilizes those amino acids that it needs to synthesize the necessary proteins required to sustain life. The amino acids not required for growth or maintenance are degraded to urea with the release of energy. The body tends to maintain a nitrogen balance wherein the intake of nitrogen from food is equal to the nitrogen excreted in a nongrowth situation. If protein is withheld from the diet, the balance can become negative.

V. STRUCTURE

Proteins are large molecules with molecular weights on the order of 10^5 or 10^6. Their structure is most readily described as primary, secondary, tertiary, and quaternary forms.

A. Primary

The primary structure of proteins is determined by the specific sequence of amino acids in the molecular chain. Amino acids contain both basic amino groups and acidic carboxyl groups, and have the general structure:

$$NH_2-\overset{\displaystyle H}{\underset{\displaystyle R}{C}}-COOH$$

There are 21 naturally occurring amino acids, of which 10 are essential for human growth and maintenance. These are listed in Table 2.

TABLE 2   Essential Amino Acids

$$NH_2-CH-COOH$$
$$|$$
$$R$$

Name	R-group
Valine	$(CH_3)_2CH-$
Leucine	$(CH_3)_2CHCH_2-$
Isoleucine	CH_3CH_2CH- $\quad\quad\quad |$ $\quad\quad\quad CH_3$
Phenylalanine	$C_6H_5CH_2-$
Threonine	CH_3CH- $\quad\quad |$ $\quad\quad OH$
Lysine	$H_2N(CH_2)_4-$
Methionine	$CH_3SCH_2CH_2-$
Tryptophan	
Arginine[a]	$HN=CNH(CH_2)_3-$
Histidine[a]	

[a]Essential only for growth.

Amino acids are joined chemically in proteins through the peptide bond when the amino group of one amino acid reacts with the carboxyl group of another amino acid:

$$
\underset{\underset{R}{|}}{\overset{\overset{H}{|}}{NH_2-C-COOH}} + \underset{\underset{H}{|}}{\overset{\overset{R'}{|}}{NH_2-C-COOH}} \rightarrow
$$

$$
\underset{\underset{R}{|}}{\overset{\overset{H}{|}}{NH_2-C-}} \overset{\overset{O}{\overset{\|}{}}}{C-O-NH} \underset{\underset{H}{|}}{\overset{\overset{R'}{|}}{-C-COOH}}
$$

The combinations possible with 21 different amino acids is virtually unlimited. The specific sequence of amino acids determines many of the fundamental properties of proteins and subsequent structure.

B. Secondary

The secondary structure of protein molecules refers to their three-dimensional spatial configuration. Again, the sequence of amino acids will influence this structure. There is debate (27) as to whether the molecule configures itself to achieve a global free energy minimum in the thermodynamic sense or whether the folding of the chain occurs along kinetically accessible pathways, which may not include the thermodynamic global minimum.

Secondary structure is the spatial arrangement that the polypeptide chain takes along its axis. This structure exhibits itself in several forms.

The α helix form such as exhibited by enzymes, globular proteins (albumins, caseinogen, gluten, etc.), and wool is shown schematically in Fig. 5.

The β pleated sheet, which is a zigzag pattern of polypeptide chains, is inelastic and typical of the protein fibroin of silk, as shown in Fig. 6.

The collagen helix, which is a stable triple helix, is yet another structure for fibrous protein.

C. Tertiary

Tertiary structure refers primarily to the bonds between active groups on the polypeptide chain. To a great extent, this determines the folding pattern of the α helix chain of globular proteins.

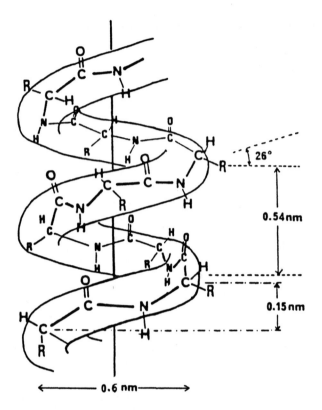

FIGURE 5   Schematic three-dimensional structure of an α helix (right-handed) (R are amino acid side chains). (From Ref. 6, used with permission.)

The types of bonds involved in tertiary structure, as illustrated in Fig. 7, are electrostatic interaction as between terminal or protruding amino and carboxyl groups, hydrogen bonding, hydrophobic interaction, dipole-dipole interaction, and covalent disulfide bonding, which is the strongest of the bonds.   The amino acid cysteine often is involved in forming disulfide linkages.   Cysteine has the following formula:

$$NH_2-CH-COOH$$
$$\qquad\ \ |$$
$$\qquad\ \ CH_2$$
$$\qquad\ \ |$$
$$\qquad\ \ SH$$

Cysteine is involved in disulfide bonding as illustrated here:

```
        ┌──S──────────S──┐
   −CY−CY−Alanine−Serine−Valine−CY
        /
        S
        /
        S
        /
  −Leucine−CY−Glycine−Serine−
```

D. Quaternary

Quaternary structure involves more than one polypeptide chain. The bonds stabilizing a multi-chain structure are similar to those considered under tertiary structure. The disulfide bond is between chains rather than intrachain bonds.

VI. FOOD PROTEINS

With the notable exceptions of myosin and glutenin, the food proteins are of the globular type. They are α helixes with the helix chain folded upon itself in a way determined by the amino acid sequence in the polypeptide chain such that the configuration in the solvent at hand represents some minimum free energy. In the manipulation of food proteins, it is important to understand two phenomena: (a) the concept of solubility and the isoelectric point, and (b) the concept of denaturation.

VII. SOLUBILITY

In solution, the formula for an amino acid is most accurately represented by the following:

$$\overset{+}{NH_3}−CH−\overset{-}{COO}$$
$$\quad\quad\ \underset{R}{|}$$

This illustrates the ionic nature of amino acids. An amino acid may be neutral, positive, or negative depending on the pH of the system. When an amino acid is neutral, it is said to be at its isoelectric point. Many properties are at a minimum or maximum at the isoelectric point. Solubility is at a minimum at the pH corresponding to the isoelectric point. Though it is not obvious why this is so, consider that while the net charge on the molecule is zero it is

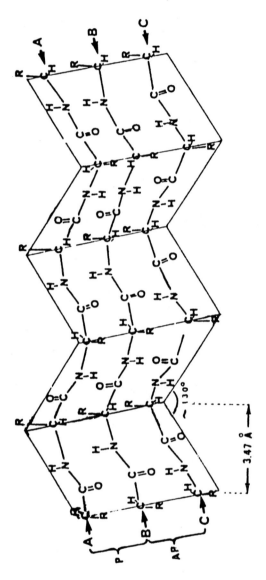

FIGURE 6 Schematic three-dimensional structure of β pleated sheets. P, two parallel β-sheet strands (between A and B chains); AP, two antiparallel β-sheet strands (between B and C chains). (From Ref. 6, used with permission.)

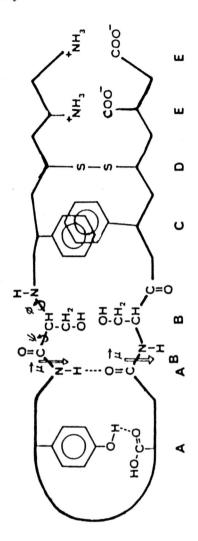

FIGURE 7 Bonds or interactions that determine secondary and tertiary structure of proteins: (A) hydrogen bond; (B) dipolar interaction; (C) hydrophobic interaction; (D) disulfide linkage, and (E) ionic interaction. (From Ref. 6, used with permission.)

still a dipole and the electrostatic attraction between neighboring molecules is enhanced, allowing them to pack closely together. Hydrogen bonding can also influence the close packing with the result that solvent molecule interaction is reduced and overall solubility decreases. Proteins that consist of a chain of amino acids behave in a similar manner. At the isoelectric point, the proteins no longer repel one another due to lack of net charges. The molecules have more opportunity to touch and aggregate, with a decrease in solubility resulting. Protein solubility in water, then, is a result of hydrogen bonding between water and protein at pH values that are not at the isoelectric point. The effect of pH on protein solubility is illustrated in Fig. 8 for the case of soybean meal protein. Note the pronounced minimum at the isoelectric point. The minimum solubility at the isoelectric point is utilized in such processes as the manufacture of cheese and soy protein isolates.

VIII. DENATURATION

Denaturation has been defined as any modification of the secondary, tertiary, or quaternary protein structure excluding cleavage of peptide bonds. In water, a protein will configure itself (fold) in such a manner as to place hydrophobic groups towards the interior of the coiled structure and place hydrophilic groups in a position to interact most fully with water molecules to stabilize the protein. The molecule is held in this configuration, in some minimum free energy state, by the bonds associated with tertiary structure.

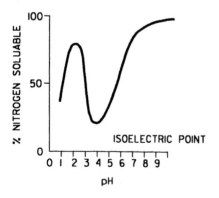

FIGURE 8   Solubility of soy protein versus pH.

Denaturation is initiated by the addition of energy through heating or acids, alkali, alcohols, and mechanical work, which all tend to disrupt the tertiary bonds and cause the molecule to uncoil. In so doing, the hydrophobic groups are exposed to the solvent, solubility is decreased, and, in the more random coil structure, aggregates of molecules are more readily formed. This is illustrated in Fig. 9 for an unspecified polypeptide chain. Taken to the extreme,

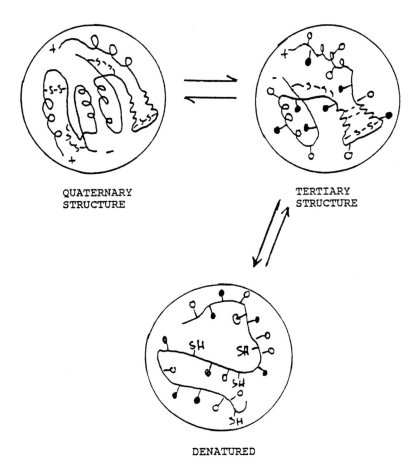

QUATERNARY
STRUCTURE

TERTIARY
STRUCTURE

DENATURED

FIGURE 9 Illustration of the denaturation of a protein molecule. Solid circles, hydrophobic sites; open circles, polar and ionic circle chains.

such as with prolonged heating, denaturation is irreversible. The cooking of an egg by boiling or frying is an everyday example of irreversible protein denaturation. In addition to coagulation, denaturation results in decrease or loss of enzymatic activity and exposure of peptide bonds for enzymatic hydrolysis. In food processing, denaturation is used to advantage in many instances to "heat set" protein to impart texture.

IX. COMPOSITION

Naturally occurring protein foods do not consist of a singular protein; they are a mixture of several different proteins. Milk is perhaps the most familiar protein food and has many components comprising its protein system. Some of these are shown in Table 3. Milk consists of two major fractions. Casein, which constitutes 80% of the total protein, is that fraction which precipitates at pH 4.6. The fraction still soluble at pH 4.6 comprises 20% of the total protein and is called "whey." Table 3 shows that there are 14 major proteins in casein and two in whey.

X. NUTRITIONAL VALUE

Protein foods are required to supply nitrogen and amino acids for the human metabolic processes culminating in the manufacture of body proteins and other nitrogen-containing compounds. Of the amino acids comprising the body proteins, nine (eight in adults) cannot be synthesized by the body and must be supplied by diet. These are called the essential amino acids (Table 2).

The nutritional value of a protein food should be judged on the basis of three factors:

TABLE 3   Major Protein Components of Milk Protein System

Casein, 80%, PPT @ pH 4.6	α_s caseins (6), 50–55%
	β caseins (3), 30–35%
	γ caseins (3), 5%
	κ caseins (2), 15%
Whey, 20%, sol. @ pH 4.6	β-lactoglobulin
	α-lactalbumin

Protein content
Digestibility
Amino acid balance

Inherent in the first factor is cost. The cost of protein varies widely even when expressed on the basis of cost per pound of protein. This is due to several factors, including availability and aesthetic appeal or desirability as a food. Some typical costs are shown in Table 4.

"Digestibility" implies that the protein must be hydrolyzed by the digestive system and reduced to the component amino acids to be of human nutritional value. The food must be devoid of any inhibitors that might reduce the rate of hydrolysis, and the protein must be in a physical form such that the digestive process is not diffusion

TABLE 4   Cost of Protein Foods on As-Is and Cost/Pound-of-Protein Basis: Retail Cost to Consumers, Informal Supermarket Survey, Minneapolis, December 1987

Source	Percent protein (as is)[a]	Price per pound (as is)	Cost per pound of protein
Beef (ground)	17	$1.49	$ 8.76
Eggs	13	0.50	3.85
Fish (cod)	18	4.55	25.29
Sardines	24	3.31	13.78
Milk (skim)	3—4	0.21	5.80
Cottage cheese	17	1.07	6.31
Peanut butter	25	1.59	6.36
Soybeans	34	0.12	0.35
Wheat flour (white)	12	0.22	1.83
Beans	22	0.51	2.32
Cornmeal	8	0.44	5.48
Rice	7	0.51	7.29

[a]Handbook no. 8, USDA.

limited. Tough meat and overcooked protein may be difficult to hydrolyze. Chemically bound amino acids, such as Maillard reaction
products with reducing sugars, are not available nutritionally. In
high protein foods, this is generally not a serious issue. In cereal
grains, browning can be a negative since it is the limiting amino acid
lysine that is bound by reaction with sugars. For example, bread
can lose 10–15% of its lysine during baking and another 5–10% during toasting (7).

For biosynthesis of proteins to occur, the necessary amino acids
must be available in the body. A well-balanced protein food contains amino acids in the proportions required by the body. There
is some difficulty in specifying what this ratio should be since many
factors need to be considered. Variables such as age, sex, quantity of protein intake, and metabolic variability determine the desired proportions of amino acids. All this is complicated by the difficulty of conducting nutritional experimentation on animals and trans-
lating to humans. For many years, the standard amino acid profile
was taken as that of whole egg protein. Animal feeding studies
have shown these to be utilized 100%, which would seem to imply an
optimum amino acid balance. It is not an unreasonable assumption
since the egg contains all the nutrients necessary to support the
life of a developing chick. However, use of whole egg protein appears to be conservative since the protein quality is beyond that
required for human nutrition with respect to some of the essential
amino acids.

More recently, attempts have been made to develop a scoring
system based on human nutritional needs (8). Using whole egg
protein as a basis, the concept of chemical score for a protein food
has been in use for some time. It is only one measure of a complicated phenomenon and, at best, gives directional and not quantitative information. It is a relatively simple concept. To use the
chemical score, it is necessary to know the amino acid composition
of both the subject protein and whole egg protein. By comparison, one finds the amino acid that is most deficient. The chemical score is then calculated expressing the amount of limiting
amino acid as a percentage of the amount in whole egg. Table 5
illustrates the method of calculation for whole wheat and defatted
soybean meal.

For whole wheat, lysine is the limiting amino acid, and the chemical score is $(2.6/6.4)100 = 40$.

For defatted soy meal, the limiting amino acid is methionine, with
a chemical score of $(1.4/5.5)100 = 25$.

What can one conclude from a comparison of chemical scores?
Unfortunately, not much. The relative chemical scores for defatted
soy flour and whole wheat would imply that wheat flour is a better

TABLE 5 Calculation of Chemical Score

		Amino acid content (g/16 gN)		
	Egg	Whole wheat	Percent deviation from egg	Defatted soybean meal	Percent deviation from egg
Phenylalanine	10.1	8.1	-20	4.46	-56
Histidine	2.4	1.9	-21	2.25	6
Tryptophan	1.2	1.6	+33	1.17	0
Leucine	8.8	6.3	-28	6.7	-24
Methionine	5.5	3.5	-36	1.4	-75
Isoleucine	6.6	4.0	-40	4.4	-33
Valine	7.4	4.3	-42	4.5	-39
Threonine	5.0	2.7	-46	3.7	-26
Lysine	6.4	2.6	-60	6.0	-6

Source: Adapted from Ref. 8, used with permission.

protein source. This is not the case at all. Soybean meal is a considerably better protein source than wheat flour.

XI. MEASURING PROTEIN QUALITY

Several methods are available for experimentally measuring quality of a specific protein food. Chemical score as illustrated above is not an experimental method and is not sufficient for several reasons: Namely, while egg protein as a basis is too conservative, the method neglects factors such as inhibitors that might be present and availability of amino acids. It also considers only one limiting amino acid, which as illustrated in the previous example can be misleading.

A. Biological Value

Biological value (BV) has been considered the most reliable test for protein quality. It actually measures the retention of nitrogen in the body by means of a nitrogen balance. This involves measurement of nitrogen intake and all nitrogen in waste streams. Biological value testing assumes a linear response with dosage. This can be checked by varying the dose over a range.

B. Net Protein Utilization

Net protein utilization (NPU) is similar to BV except it is based on young rats, with change in body protein being determined by carcass analysis rather than nitrogen balance.

C. Protein Efficiency Ratio

Protein efficiency ratio (PER) consists of feeding weanling rats a diet containing 10% of the test protein for a period of 28 days. The uncorrected PER is the weight gain divided by the protein eaten. A control group is also fed a diet containing casein, and the uncorrected PER is corrected to an assumed PER of 2.5 for casein. Although PER is the most widely used test, it does suffer from some serious deficiencies (11):

Since the calculation is based on weight gain only, PER minimizes the value of protein required for maintenance. It penalizes poor quality proteins.
The level of protein chosen is too high for good quality proteins, so that PER underestimates the quality of the best proteins.
The use of casein as a standard and the attempt to correct values due to rat and environmental variability are not always adequate.

PER testing is less expensive than BV. The first sample for PER is $850 and includes a casein control. Subsequent samples are $425. The first sample for BV is $1,500 and subsequent samples are $1,075 each—as of 1987 (10).

The PERs of some common food proteins are listed in Table 6.

Some interesting observations can be made upon examining the data in Table 6. Consider first that the PER of soybean meal (1.9) is greater than that of soy concentrate (1.7), which is greater than that of soy isolate (1.3). This is a direct result of protein fractionation that occurs in the processing of soybean meal to manufacture concentrates and isolates. The decrease in PER is due to preferential loss of protein fractions high in methionine-cysteine in both the concentrate and isolate processes. In effect, some of the nutritional value is lost in the process of removing unwanted carbohydrates from the protein.

The data on bread and toast illustrate another form of processing that results in nutritive loss, but in this instance it occurs in

TABLE 6   Protein Efficiency Ratios of Common
Protein Foods

Whole egg protein	3.8
Beef	3.2
Milk (whole)	2.5
Soybean meal (toasted)	1.9
Soy concentrate	1.7
Soy isolate	1.3
Peanut	1.7
Corn	1.2
Bread	1.0
Light toast	0.64
Dark toast	0.32

Source: Adapted from Ref. 22, used with per-
mission.
Thomas Richardson and J. W. Finley, Chemical
Changes in Food During Processing, p. 456,
AVI, 1985, Westport, Ct.

the home. The PER of bread (1.0) is greater than light toast (0.64),
which is greater than dark toast (0.32). This decline is the result of
chemical binding of lysine due to the Maillard reaction between reducing sugar and the ϵ amino acids (primarily lysine) in the protein of wheat. A certain amount of browning of bread crust due to
the Maillard reaction is desirable for flavor and color development
during baking. However, in the toasting process in the home the
interior portion of the bread is subjected to much higher temperatures than in the baking process, with the result that lysine is effectively consumed in the formation of Maillard browning products.
Maillard reaction renders the lysine unavailable for human nutritional purposes and depletes the bread of the amino acid that happens to be limiting.

XII. CARBOHYDRATES

Carbohydrates, or hydrates of carbon, have the general formula
$C_x(H_2O)_y$, where x and y are whole numbers. Plants synthesize
carbohydrates from carbon dioxide and water, with energy supplied

by sunlight in the process called "photosynthesis." Animals are not capable of synthesizing carbohydrates and are thus dependent on plants for glucose.

A. Sugars

The simpler carbohydrates are called "sugars" and are typically crystalline solids, which dissolve in water to give a sweet taste and consist of monosaccharides and disaccharides.

B. Oligosaccharides

Oligosaccharides are water-soluble polymers of a few condensed monosaccharides (typically two to 10). Disaccharides are also oligosaccharides.

C. Polysaccharides

Polysaccharides are high molecular weight polymers of glycosidic sugar units. They are noncrystalline and generally insoluble in water.

The basic monomeric units of the carbohydrates are glucose, fructose, and galactose, shown in their conventional ring configuration.

D-glucose (dextrose):

D-fructose (levulose):

D-galactose:

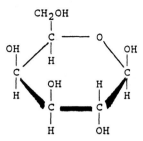

In terms of these specific monomeric units, it is useful to tabu-
late some common food carbohydrates along with the degree of poly-
merization (Table 7).

It is interesting to contrast the structures of two abundant polysaccharides: starch and cellulose. Starch is a mixture of two substances: amylose and amylopectin. Amylose is a linear polymer consisting of glucose units connected with α linkages as depicted in
Fig. 10. Amylopectin is much more complex in that it too consists
of glucose units connected with α linkages, but the polymeric chain
has branch points and side chains with the branching occurring between the reducing group at the end of one chain and the primary
alcohol group of another. There is generally a multiplicity of branches with complex three-dimensional structures. Starch, whether it be
amylose or amylopectin, is hydrolyzed by the enzymes α and β amylase, which are present in plants and animals. In the case of amylose, the amylase enzyme completely hydrolyzes it to maltose and
glucose. This occurs because amylase splits off pairs of glucose
units from the free ends of the polymeric chain. Amylopectin, by
contrast, is a nonlinear chain, and amylase can function only until
it encounters a branch point. It is then unable to split off a glucose pair, and hydrolysis ceases at that branch point.

The polymeric pieces that are left in this process are called "limit
dextrins." Complete hydrolysis of starch is possible by enzymatic
action of glucoamylases (26).

Cellulose is similar in structure to starch except for the β-linkage
rather than α-linkage between glucose units. Amylase is not selective for this bond, and cellulose is indigestible by humans and
nonruminant animals. Cellulose can be hydrolyzed by strong acid
or by the enzyme cellulase. Several enzymes are included in the
cellulase complex. The hydrolysis can follow several paths (18).
In one scheme, an endoglucanase cleaves the cellulose chain, exo-
glucanase can then clip cellobiose (two glucose units) from the free
ends. A third enzyme, cellabiase, then hydrolyzes the cellobiose to
glucose. Considerable effort continues to be directed at developing

TABLE 7 Food Carbohydrates

	Monomer units	Degree of polymerization	Common source	Digesti-bility
Disaccharides				
Sucrose	D-Glucose, D-fructose	2	Sugar cane, sugar beets, fruits	H
Maltose	D-Glucose	2	Starch syrups, malt, honey	H
Lactose	D-Galactose, D-glucose	2	Milk	H/M
Oligosaccharides		(2–10)		
Raffinose	Galactose, glucose, fructose	3	Legume seeds	M
Stachyose	Galactose (2), glucose, fructose	4		
Polysaccharides				
Starches				
Amylose	Glucose	70–350	Cereals, roots, tubers, legumes	H
Amylopectin	Glucose	100,000	Same	H
Glycogen	Glucose	5,000	Liver, tissue	H
Cellulose	Glucose	100,000	Plant cell walls and fiber	N

H, high; M, medium; N, not in humans.

economically feasible processes to convert cellulose to glucose and ultimately to ferment this to alcohol.

XIII. SUGARS

A. Sweetness

Sugars do provide nutritional value and are utilized primarily for their sweetening properties in food. From about 1925, consumption of cane and beet sugar had been relatively stable at 100 lb/capita through 1977. In the late 1970s, consumption of sugar began a decline as the consumption of corn sweetener increased from 11.5 lb/capita in 1960 to 58 lb/capita in 1984. By 1984, cane/beet sugar consumption had declined to 67.5 lb/capita. The development of high fructose corn syrup (HFCS) and its subsequent use in soft drinks as a replacement for sucrose was a major factor in this shift in consumption patterns (1).

Current trends would indicate that a large segment of the population would be quite content if sugar had no nutritional (caloric) value. The effort being expended to develop nonnutritive artificial sweeteners would seem to support such a premise. Consumption of nonnutritive sweeteners has grown from 2.9 lb/capita in 1950 to 15.8 lb/capita in 1984. Of this latest figure, saccharin accounted for 63% and Aspartame for 37% (15). Interestingly, approval for use of Aspartame in carbonated beverages was granted in 1983. Sold under the trade name Nutrasweet, sales of Aspartame were well over $700 million in 1986 (2).

It is not clearly understood why some substances impact a sweet taste and others do not. There have been many attempts to explain this—none completely successful. A popular theory that has a stereochemical basis suggests that there is a taste triangle on the molecular level, as depicted in Fig. 11.

Presumably, this structure forms reciprocal hydrogen bonds with a similarly configured receptor site in the taste buds, with the response being a sweet taste (20).

Fructose has such a configuration and is indeed the sweetest of the naturally occurring, nutritive sweeteners. The theory's efficacy has been demonstrated for pseudofructopyranose, which has a carbon atom substituted for the oxygen atom in fructopyranose (fructose) (24). It has also been reported that trichlorogalactosucrose, a noncaloric sweetener, which has two taste triangles, is 2,000 times as sweet as the sugar from which it is derived (4).

Sensory analysis is generally the method used to determine sweetness. It is difficult to measure because sweetness is a subjective reponse detected by receptors in the tongue and is influenced by other factors. For example, if one compares the relative sweetness

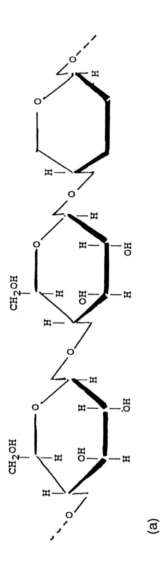

(a)

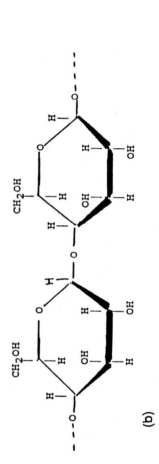

(b)

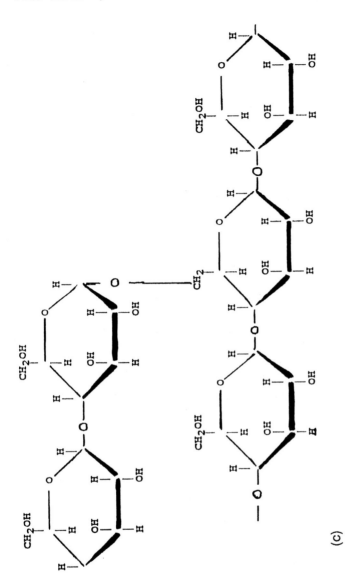

FIGURE 10 Polysaccharide structure. (a) Cellulose, glucose polymer with β linkage; (b) amylose, glucose polymer with α linkage; and (c) amylopectin.

(c)

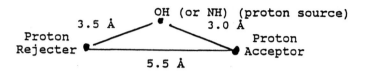

FIGURE 11   Taste triangle.

of several sugars, the concentration of the subject sugar solutions can affect the outcome. Results obtained at the threshold level of sweetness may not hold true at levels of normal consumption. Temperature determines the isomeric form of fructose, and some forms are sweeter than others. Some generalization can be made, however. The common sweeteners—glucose, sucrose, and fructose—are compared in Table 8. Although the exact numbers can be disputed, the directional trends are essentially correct relative to one another and reproducible. Taste panels are capable of reproducing rankings such as these. It is interesting to note the relative sweetness of Aspartame and saccharin. Even if nutritive carriers, such as dextrins or polydextrose, must be used on consumer applications, such as table sugar replacement, the caloric reduction is quite significant since these carriers are only about 25% the caloric intensity of sugar. In applications, such as soft drinks, the full caloric benefit can be achieved since carriers are unnecessary.

B. Viscosity Effect

The perception of sweetness can be affected by physical factors not related to the chemical structure of a sweetener. The effect of viscosity or perceived sweetness has been quantified both experimentally and in terms of diffusion theory (13).

The key assumption in explaining the viscosity/sweetness interaction is that the chemical reactions occurring in the taste buds are fast compared to the rate of diffusion of sweetener through the carrier medium to the taste buds. The process is, thus, diffusion-limited, and the taste intensity is proportional to the flux of sweetener reaching the tongue.

From diffusion theory, we see that taste intensity $\propto$ (amount of sweetener reaching tongue) $\propto CD^{1/2}$ where C is concentration in the mouth, and D is the diffusion coefficient of sweetener in the carrier. Agreement between theory and experimental data for sucrose in tomato solids showed a correlation coefficient of 0.91.

TABLE 8   Comparison of Common Sweeteners

	STRUCTURE	RELATIVE SWEETNESS SCORE
GLUCOSE		75
SUCROSE		100
FRUCTOSE		123
ASPARTAME®		20,000
SACCHARIN		30,000

C. Hygroscopicity

Sugars and dextrins are hygroscopic. The amount of water that
can be absorbed from the atmosphere varies among sugar types.
Invert sugar and fructose will absorb four to five times as much
water as sucrose. Glucose is considerably more hygroscopic than
crystalline sucrose, as indicated in Fig. 12. This can have significant consequences in a manufacturing operation:

Moisture pickup in stored ingredients can affect batch formulas that
    are mixed on a weight basis.
In humid environments, products can become sticky and difficult to
    handle and convey.
Hygroscopic products must have adequate packaging protection to
    maintain shelf life.

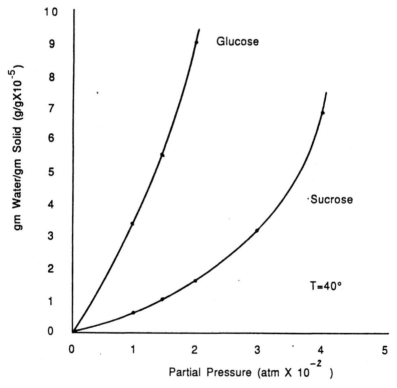

FIGURE 12   Sorption isotherms for glucose and sucrose at 40°C.
(From Ref. 23, used with permission.)

D. Browning Reactions

The various browning reactions of sugars, whether they be carmelization or Maillard (sugar/amine), occur frequently in food processing in either a controlled or uncontrolled manner. The Maillard reaction (which is actually a series of several reactions) is the most
frequently encountered in food processing operations. The initiating
reaction is between the amino groups of protein or amino acids and
the carbonyl group of a reducing sugar. The end products are
generally brown. Intermediate products in the reaction sequence
often have distinctive and desirable flavors.

   Maillard reactions are very complex and are responsible for many
flavor and color compounds depending on initial reactants and conditions such as pH, temperature, and water activity.

   The required amino group can come from free amino acids or proteins containing amino acids with a free amino group (dibasic). Lysine, arginine, and histidine are essential amino acids in this latter
group. Reducing sugars have the ability to form an open chain
structure and expose the aldehyde group. This includes almost all
sugars, except the disaccharide sucrose and the sugar alcohols
(sorbitol, mannitol, xylitol, etc.). Sucrose must hydrolyze first
to glucose and fructose to enter into Maillard reactions.

   It is thought that the initial reaction rate between the reducing
sugar and amino group is determined by the ability of the sugar's
ring structure to open through mutarotation to an open chain re-
ducible form. This is illustrated in Fig. 13 for D-glucose.

   The equilibrium can be affected by pH and the presence of met-
als such as iron or copper. It has been suggested that browning
rates are proportional to mutarotation rates with some evidence presented for support (14).

   The complex Maillard sequence is summarized in Fig. 14 in general form and Fig. 15 specifically for glucose.

   The reducing sugar and amine react reversibly to give a glycosyl-
amine, which undergoes Amadori rearrangement. The Amadori compound can then follow two paths depending upon pH. For slightly
acidic conditions, the sequence ends with a Furan derivative. In
less acid medium, the sequence ends with dicarbonyl compounds,
which can form cyclic flavor compounds. Considerably more detail
can be found in standard reference texts (5,12,25).

   The cyclic compounds maltol and isomaltol are "caramel-like" in
flavor and contribute to the flavor of baked products.

   Maillard reaction products contribute significantly to the desirable flavor of products such as breakfast cereals, cookies, breads,
cakes, nuts, coffee, beer, and the like.

   However, in some products that are of a delicate flavor, such as
dry nonfat milk solids, browning reactions can be undesirable in

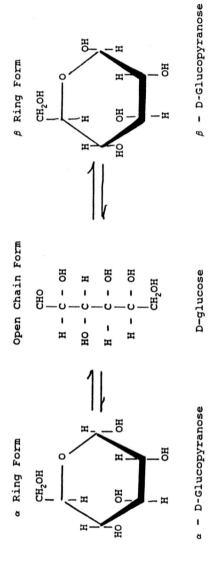

FIGURE 13 Equilibrium between ring and open chain for reducing sugars.

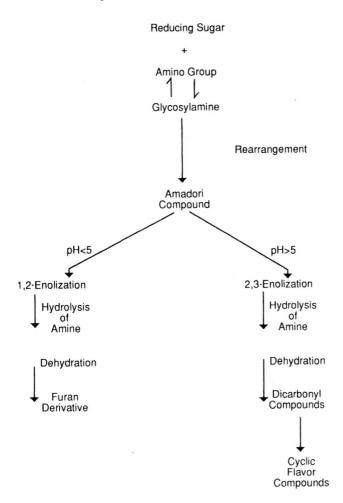

FIGURE 14   General Maillard reaction sequence.

that the product will develop flavors and colors that were not intended.   Shelf life could be significantly reduced.   Care must be
taken during processing to avoid conditions that promote browning
reactions if browning would be detrimental to the product.   Several
precautions can be taken:

Maintain low temperatures during processing and subsequent stor-
    age.

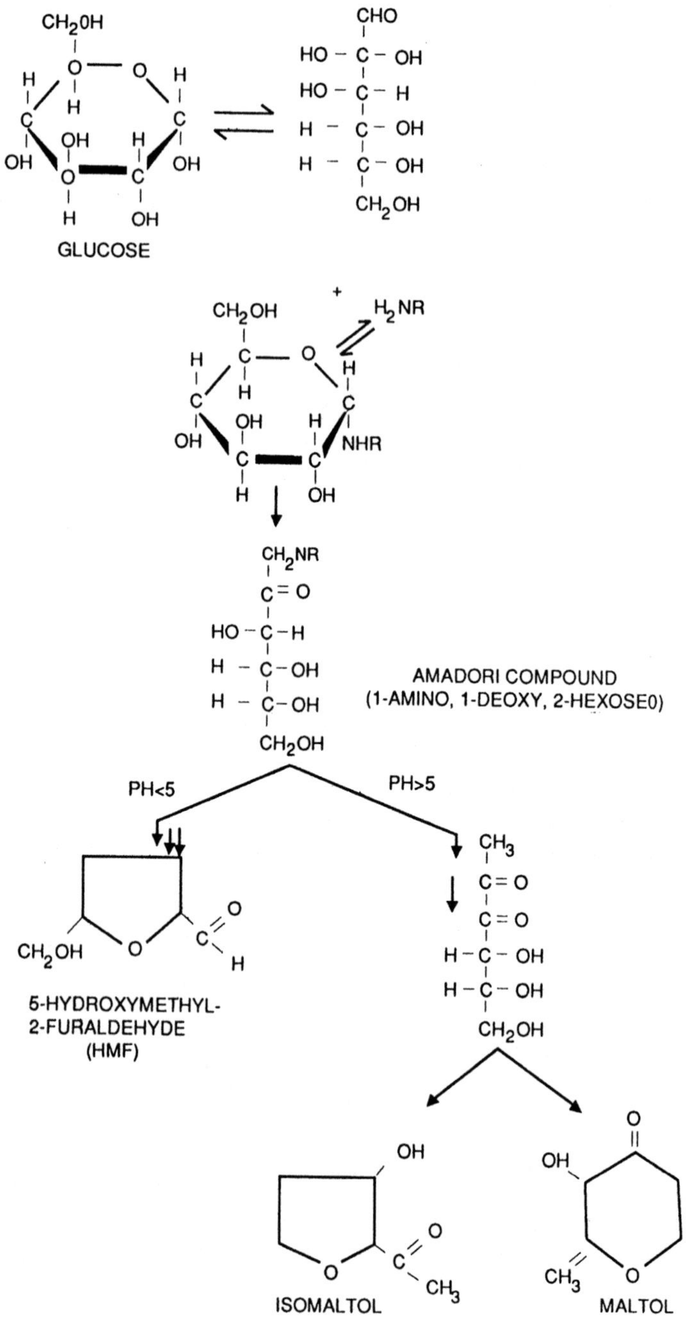

FIGURE 15 Maillard reaction scheme for glucose.

For sensitive products, temperature-regulated storage may be necessary, especially in warm climates.

In designing a process or plant, these factors should be considered so that exposure to high temperatures is minimized, especially in postprocess handling and surge systems.

E. Starch

Starch is widely used throughout the food processing industry. It occurs naturally in various plants in the form of small granules whose shape and size are characteristic of the particular plant. Starch is not soluble in cold water, and it will only absorb about 30—45% water unless it is heated. The resistance to moisture absorption is due to the intermolecular hydrogen bonds between linear amylose chains or between branches of amylopectin chains. As heat is applied, the bonds are weakened and the granules begin to absorb water and swell. The temperature at which the hydrogen bonds begin to weaken and the starch granules begin to swell rapidly is caled the "initial gelatinization temperature." The onset of gelatinization can be observed through a loss of birefringence (a characteristic cross appears on the ungelatinized granule when viewed under polarized light; the cross disappears where gelatinization occurs) due to the loss of orderly molecular orientation. Gelatinization occurs over a temperature range due to variation in the size, shape, and molecular configuration within the individual granules. Starches from different sources exhibit differing gelatinization temperature ranges. Wheat starch will gelatinize at 58—64°C, rice at 68—78°C, and corn at 62—70°C (6).

Because of the hydrogen bonding between amylose and amylopectin chains, uncooked (ungelatinized) starch is difficult to digest. The enzymes in the digestive juices cannot easily penetrate the starch granules to convert the starch to maltose. In a sense, the digestive process becomes diffusion limited. Cooking or gelatinization alleviates this problem by breaking down and softening the relatively impermeable structure.

F. Viscosity of Starch Solutions

Starch solutions of varying concentrations are used extensively in food processing. Starch can be used as a simple thickener in soups and gravies in which it can exhibit shear thinning properties, or it can be converted to a dough that exhibits viscoelastic properties. The interaction between water and starch is important in determining its rheological behavior. Thus, additions that change water activity dramatically, such as high salt or sugar concentrations, can retard gelatinization and viscosity development.

Because of its highly branched structure, hydrogen bonding in amylopectin is limited and its solutions tend to remain relatively clear and fluid. This is an important attribute in product formulation, and it has made waxy maize starch, which is 99% amylopectin, a popular ingredient.

As starch granules absorb water, the molecules occupy an increasing volume. Viscosity is highly dependent on concentration, and, in fact, there are distinct differences between dilute and concentrated starch solutions. This is explained on the basis of entanglement of the starch polymer chains once a critical concentration has been reached (3).

REFERENCES

1. Anonymous, *Food Technology*, p. 112 (1986).
2. Anonymous, *Manufacturing Week*, Jan. 11, p. 39 (1988).
3. Aspinall, G. O., ed., *The Polysaccharides*, Vol. 1, p. 259, Academic Press, Inc., New York (1982).
4. Emsley, J., *New Scientist*, October, p. 22 (1985).
5. Feather, M. S., *Chemical Changes in Food During Processing*, T. Richardson and J. W. Finley, eds., AVI, Westport, Connecticut, pp. 289 ff., (1985).
6. Fennema, O. R., ed., *Food Chemistry*, 2nd ed., Marcel Dekker, Inc., New York (1985).
7. Fox, B. A., and A. G. Cameron, *Food Science—A Chemical Approach*, University of London Press, London, p. 280 (1972).
8. Harper, A. E., *Food Proteins*, J. R. Whitaker and S. R. Tannenbaum, eds., AVI, Westport, Connecticut, p. 112 (1977).
9. Halpin, B. E., and C. Y. Lee, *Journal of Food Science*, 52: 1002–1005 (1987).
10. Hazelton Labs., *Verbal Communication*, Madison, Wisconsin, December (1987).
11. Hegsted, D. M. *Food Proteins*, J. R. Whitaker and S. R. Tannenbaum, eds., AVI, Westport, Connecticut, p. 354 (1977).
12. Hodge, J. E., and E. M. Osman, *Principles of Food Science*, O. R. Fennema, ed., Marcel Dekker, Inc., New York (1976).
13. Kokini, Josef L., *Food Technology*, p. 86 ff., November (1985).
14. Labuza, T. P., and M. K. Schmidl, *Role of Chemistry in the Quality of Processed Food*, O. R. Fennema, W. Chang, and C. Lii, eds., Food and Nutrition Press, Inc., Westport, Connecticut, p. 79 (1986).

15. Lecos, C., *FDA Consumer*, 19:25 (1985).
16. Luna, J. A., R. L. Garrote, and J. A. Bressau, *Journal of Food Science*, 51:141—145 (1986).
17. Lund, Darrly B., *Information Bulletin #87-3*, National Food Processors Association, pp. 3—5 (1987).
18. Montencourt, B. S., *Wood and Agricultural Residues*, E. J. Soltes, ed., Academic Press, New York, pp. 271 ff. (1983).
19. Pomeranz, Y., *Functional Properties of Food Components*, Academic Press, Orlando, Florida (1985).
20. Shallenberger, R. S., and T. E. Acree, *J. Agric. Food Chem.*, 17:701 (1969).
21. Singh, R. P., and J. H. Wells, *Int. J. Refrig.*, 10:296—300 (1987).
22. Smith, A. K., and S. J. Circle, *Soybeans: Chemistry and Technology*, AVI, Westport, Connecticut, p. 112 (1980).
23. Smith, D. C., C. H. Mannheim, and S. G. Gilbert, *Journal of Food Science*, 48:1051 (1981).
24. Suami, T., *Chemistry Letters*, p. 719 (1985).
25. Waller, G. R., and M. S. Feather, *The Maillard Reaction in Foods and Nutrition*, ACS Symposium Series 215, ACS, Washington, D.C. (1983).
26. Whistler, R. L., and E. F. Paechall, *Starch: Chemistry and Technology*, Vol. 1, Academic Press, New York (1965).
27. Whitaker, J. R., and S. R. Tannenbaum, *Food Proteins*, AVI, Westport, Connecticut, pp. 2—5 (1977).
28. Williams, D. C., M. H. Lim, A. O. Chen, R. Pangborn, and J. R. Whitaker, *Food Technology*, pp. 130—140 (1986).

5
Soy Complex

Much has been written about the world shortage of protein for human consumption. Vegetable proteins have been touted as a potential solution to the problem. These have been available in quantity for several years, and much effort has been expended to render them closer to meat in organoleptic properties. The failure to alleviate the world protein shortage is more political and, to some extent, a result of food prejudices than due to any failure of the food industry to develop the technology necessary to provide wholesome, economical alternatives to meat protein.

Rather than becoming involved in the socioeconomic debate, we examine here the industries and technologies that evolved from one particular vegetable protein source: the soybean. An in-depth examination of the soy complex and the multitude of technologies involved provides some insight into the more practical applications of engineering principles to food processing operations.

I. ORIGINS AND HISTORY

Soybeans are thought to have originated in Asia. The earliest written record is of Chinese origin dating back to about 2800 B.C. It is not known how soybeans were first utilized for food, but the preponderance of fermented foods based on the soybean would imply that this was a preferred form. Products such as shoyu (soy sauce) and miso date back over 2,000 years. Nonfermented products such as soybean milk and tofu (soybean curd) have been available for almost as long.

The growth of the soybean industry in the United States was not based on products like tofu and shoyu. Soybeans contain 18% oil, and it is this oil that was the "reason for being" in the U.S. agricultural system. Soybean oil was imported from Manchuria in the early 1900s to supplement a shortage of cottonseed and flax oil. Imported beans were processed in 1911, and the first domestic beans were processed in 1915 (6). By 1922, the soybean crop in the United States was 4 million bushels. In 1982, the U.S. produced approximately 1.57 billion bushels. Soybean oil and meal have both become very important economically to the United States and the world.

Today, many familiar products are produced from soybeans as part of the soy complex. These include cooking oils and fats, salad oil, salad dressings, margarine, lecithin, soy protein isolates and concentrates, texturized meat analogs, meat extenders, and animal feed. There are numerous food products that contain derivatives of soybean oil and protein.

In addition to its desirability as an oilseed, the soybean is also highly valued for its high protein content. The approximate composition of soybean is given in Table 1.

There are several reasons for the soybean's prominence as a prime source of vegetable protein. The principal reasons are as follows:

Abundance
High protein content
Good quality protein
Historical use as food
Inexpensive

TABLE 1   Composition of Soybean (as is basis)

Component	Percentage (%)
Protein (Nx6.25)	37.2
Lipid	18.6
Carbohydrate	28.0
Ash	4.6
Hull (fiber)	4.6
Moisture	7.0
Total	100.0

II. PROCESSING

The ancient Chinese learned that while soybeans could be soaked in water and cooked, more interesting and useful products could be manufactured from the constituents of the soybean. Soy milk and tofu are two prodcuts manufactured from soybeans that have been consumed in Asia since 200 B.C. The processes were developed by trial and error without knowledge of the physical and chemical reasons for the various processing steps. This is a truly remarkable example of food processing considering the level of knowledge at that time. The processes for the manufacture of soy milk and tofu are attempts to refine a raw ingredient by extracting the desirable fat and protein fractions from the soybean. The concept of a value-added product is not new. Soy milk is an intermediate in the tofu process. The process is depicted schematically in Fig. 1. Soybeans are soaked in water for several hours to soften them. Grinding, in excess water, is necessary to increase surface area and reduce critical dimensions to enhance the heat and mass transfer operations in subsequent steps. Heating or cooking the mash serves several purposes. Volatile "beany" flavors are stripped off. Heating also inactivates proteinase inhibitors (trypsin inhibitors) and destroys hemagglutinin (plant proteins that agglutinate blood) activity. Filtration removes insoluble carbohydrate and hull material from the protein/oil emulsion. The filtrate is soy milk, which is similar in appearance and protein content to cow's milk. As with all soy-based products, the amino acid profile is deficient in methionine. Unlike cow's milk, soy milk contains no lactose. The PER (protein efficiency ratio) of soy milk is reported as 2.0, with cow's milk at 2.5 (6). Addition of methionine would increase the PER to that of cow's milk. Even at a PER of 2.0, soy milk is a good protein source. The PER of the soy milk will be reduced significantly if heat treatment is not adequate to inactivate trypsin inhibitors. Soy milk can be marketed itself or further processed to convert it to tofu. To accomplish this, calcium sulfate is added to the soy milk at 2–4% of the weight of the soybeans used in the batch. The calcium causes the protein to aggregate through calcium bridging and eventually to precipitate. The oil, since it is part of an emulsion with the protein, precipitates as well. The resulting curd is gel-like in structure. Further refinement of the soy milk occurs during this process of coagulation. The whey, which is expelled, contains soluble carbohydrates such as sucrose, stachyose, and raffinose. Additionally, hemagglutinins tend to be soluble in the whey. The whey is easily removed by decanting.

There is some protein fractionation in the coagulation process. A certain fraction of the protein is soluble in the whey. This alters the amino acid balance negatively since the PER of tofu is 1.8 compared to soy milk at 2.0.

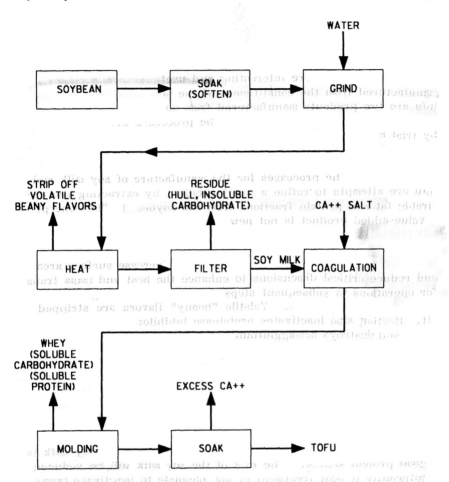

FIGURE 1   Tofu manufacture.

The removal of stachyose and raffinose is important since these can cause gastrointestinal discomfort and flatulence because they are not readily hydrolyzed in the digestive process but rather are fermented in the large intestine.

III. ECONOMICS

The point was made previously that one reason for the prominence of soybean as a protein source is its low cost. How much less expensive is soy protein when compared to traditional sources? How does one make a fair comparison?

It would seem reasonable to compare a soy-based product to the product it is intended to replace. Tofu comes close to cheese in protein and calorie content. It has less fat, which is also highly unsaturated, as compared to butterfat. With the current thinking relative to dietary fat, consumption of tofu would seem to have an advantage both in fat content and degree of saturation over cheese. The basis in this case should be 2% butterfat milk so that the final products are close in protein/fat composition. In the manufacture of cheese, about 85% of the protein precipitates and is recovered. All of the fat is recovered.

The average composition of 2% milk is as follows:

Water, 89.3%
Fat, 2.0%
Protein, 3.3%
Lactose, 4.7%
Ash, 0.7%

The yield of cheese from 100 lb of 2% milk can be estimated as follows: 2.0 (fat) + 0.85(3.3) (protein) = 4.81 lb solids as cheese.

Similarly, the yield in the tofu process can be estimated from the composition of tofu and soybean:

	Tofu (moisture-free basis) (%)	Soybean (%)
Protein	50	37
Fat	29	18.6
Carbohydrate	16	28
Ash	5	4.6
Hull	—	4.6
Water	—	7

The same assumption, that all the fat is recovered, is valid with tofu as well as cheese. Basis: 100 lb soybean, 18.6 lbs of fat = 29% of tofu; yield tofu = 18.6/0.29 = 64 lb/100 lbs soybean.

The current prices for soybeans and 2% milk are as follows: soybeans = 0.104/lb ($6.20 bushel); 2% milk = 0.14/lb ($1.15/gal wholesale). The comparative cost of tofu and cheese are then as follows: cheese = $2.91/lb of solids; tofu = $0.16/lb of solids.

The difference is staggering. The cow must convert plant protein to milk protein and is not as efficient a nitrogen fixer as the lowly *Rhizobium japonica*, which is so effective in soybeans. The cow is considerably higher than soybean on the food chain, and the cost of producing a pound of milk protein is several times more than the cost of producing a pound of soy protein. Price supports for dairy products complicate the picture. Protein foods are not judged solely on the basis of cost per unit of protein. Organoleptic and social factors such as economic status, tradition, aesthetics, and prejudice are quite important in the ultimate selection of protein foods.

Tofu is beginning to receive more and more attention in the Western world as the cost of traditional protein sources continues to rise and as people become more concerned with the food/health relationship. As new uses are found and consumer interest increases, the manufacture of tofu will be less of a cottage industry as productivity measures are brought to bear. Batch processes will be made continuous, and considerable attention will be focused on quality.

IV. MANUFACTURING CONSIDERATIONS

Batch processing, such as the heating step, can often lead to variability with the development of off-flavors due to excessive heating or nutritional losses due to inadequate heating for inactivation of trypsin inhibitors. Control and consistency are the key ideas, and these can best be accomplished in a properly designed heat exchanger rather than a batch heating vessel. One of the best designs for heat-sensitive materials such as milk is the plate heat exchanger. In principle, the plate heat exchanger has a very high heat transfer rate so that product temperature ramps up quickly and heating times can be greatly reduced compared to batch heating. The product can also be cooled rapidly in the same manner so that carmelization and degradation reactions are greatly reduced. Plate heat exchangers are also easily controlled automatically so that uniformity of heat treatment over time is reasonably straightforward to sustain. For the case of soy milk, consider the following situation.

Example

A plate heat exchanger, shown schematically in Fig. 2, is to be used to inactivate trypsin inhibitor in soy milk by heating it from 70°F to 175°F and then immediately cooling to 70°F. Heating fluid is available at 205°F, and cooling water is available at 45°F. Sixty

Batch vs Continuous

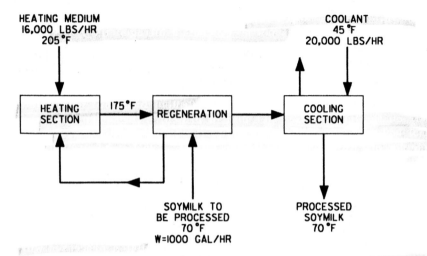

FIGURE 2   Plate heat exchanger configuration with regeneration.

percent of the heating will be done in the regeneration section, where cold feed is preheated by soy milk at 175° coming from the heating section. The heat exchanger uses rectangular plates with the dimensions 4 ft high × 1.5 ft wide with an effective surface area of 5.71 ft². The plate gap is 0.06 in. Overall heat transfer coefficients are 200 BTU/hr ft² °f in the regeneration section and 500 BTU/hr ft² °F in both the heating and cooling section. Flow rates are given in Fig. 2. Assume that soy milk has a density of 64 lb/ft³ and a specific heat of 0.91 BTU/lb °F. The heating and cooling fluids have specific heats of 1.0 BTU/lb °F.

Determine the number of plates required in each section of the heat exchanger and the residence time in the heating and regeneration section.

Energy Balance

Q_t = Total heating load

$$= \left(1,000 \ \frac{gal}{hr}\right)\left(\frac{1}{7.48} \ \frac{ft^3}{gal}\right)\left(64 \ \frac{lb}{ft^3}\right)\left(0.9 \ \frac{BTU}{lb \ °F}\right)(175 - 70)°F$$

$$= 8.085 \times 10^5 \ BTU/hr$$

Regeneration

Sixty percent of heat transfer to soy milk occurs in this section by design:

$Q_r = 0.6Q_t = 4.851 \times 10^5$ BTU/hr

Also, the temperature rise of the soy milk is 60% of the total:

ΔT cold $= 0.6 \times 105 = 63°F$

Then, the cold fluid leaves the regeneration section at $63 + 70 = 133°F$.
For the hot fluid, which enters at 175°F,

$Q = 0.6Q_t = 4.851 \times 10^5 = (8,556 \text{ lb/hr})(0.9)(175 - T)$

$T = 112°F$

175°F → ┌─────────────┐ → 112°F
        │ Regeneration│
133°F ← └─┐   │
          └───────────┘
        ↑ 1,000 gal/hr 70°F

$Q_r = U_r A_r \Delta T_r$

or

$A_r = 4.851 \times 10^5/(200 \times 42) = 58 \text{ ft}^2$

where $\Delta T_r = 42°F$ is safely used because it is the same value at both ends. Plates $= 58/5.71 = 10.16 = 11$,.

Heating Section

$Q_H = (0.4) \, Q_t = 3.234 \times 10^5$ BTU/hr

Heating fluid ΔT, $Q = 3.234 \times 10^5 = (16000)(1) \, \Delta T$, $\Delta T = 20.2$.
$T_{hot, \, out} = 205 - 20.2 = 185°F$.

          ┌───────┐ ← 205°
133°F → │ Heating│ → 175°
          └───┬───┘
            ↓
          185°

In heat exchangers with changing ΔT values, it is safe to use the logarithmetic mean:

$$\Delta T_{LM} = \frac{(185 - 133) - (205 - 175)}{\ln (52/30)} = 40°$$

$$A_H = \frac{QH}{U_h \Delta T_{LM}} = \frac{3.234 \times 10^5}{(500)\ (40)} = 16.17 \text{ ft}^2$$

$$\text{Plates} = \frac{16.17}{5.71} = 2.83 = 3 \text{ plates}$$

Cooling Section

$$Q_C = (1,000)\frac{64}{7.48}(0.9)(112 - 70) = 3.234 \times 10^5 \text{ BTU/hr}$$

$$\text{Coolant temperature rise} = \Delta T_C = \frac{3.234 \times 10^5}{20000} = 16°F$$

$$T_{coolant,\ out} = 45 + 16 = 61°F$$

112° → ┌──────────┐ ← 45°
        │ Cooling │
61° ←   └──────────┘ → 70°

$$\Delta T_{LM} = \frac{(112 - 61) - (70 - 45)}{\ln (51/25)} = 36.62$$

$$A_C = \frac{3.234 \times 10^5}{(500)(36.62)} = 17.66$$

$$\text{Plates} = 17.66/5.71 = 3.09 \simeq 3 \text{ plates}$$

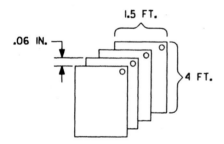

STRUCTURE 1   Residence time in heating and regeneration.

Plates = 3 (heating) + 11 (regeneration) = 14
Soy milk passes through 7 gaps
Distance over 1 plate = 4 ft
Total distance = 7 × 4 = 28 ft
Cross-section normal to flow = 1.5 ft (0.06/12) = 0.0075 ft^2
Linear velocity = 1,000/(3600)(7.48)(0.0075) ft/sec
Residence time = 28 ft/4.95 ft/sec = 5.66 sec

How does this compare to the batch process in an agitated jacketed vessel? It is clear that the residence time will be longer in a batch vessel, but is it significant enough to justify a relatively expensive and sophisticated plate heat exchanger?

The analysis of unsteady state heat transfer in jacketed agitated vessels requires an energy balance on the contents of the vessel as depicted below:

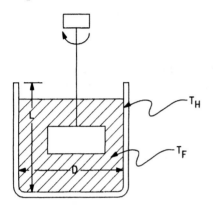

STRUCTURE 2   Jacketed, agitated vessel.

$$WC_p \frac{dT}{d\Theta} = UA (T_H - T_f) \tag{1}$$

where

T_f = temperature of fluid
T_H = heating medium temperature
C_p = specific heat of fluid being heated
W = mass of fluid in vessel
U = overall heat transfer coefficient
A = area for heat transfer

The contents are assumed to be uniformly mixed.

This equation is valid as long as T_H is constant. This is essentially correct if the heating medium is condensing steam. Then, with the initial condition that $T_f = T_0$ when $\Theta = 0$, the solution to the differential equation is:

$$\frac{UA\Theta}{WC_p} = \ln\left(\frac{T_H - T_0}{T_H - T_f}\right) \tag{2}$$

and either T_f or T_H can be calculated when the other is given. The problem then becomes one of obtaining appropriate values for U. There are essentially three choices:

1. Find U experimentally.
2. Use literature values for similar geometry and dynamic conditions.
3. Calculate U from correlations for agitated vessels, such as Chilton's (see any standard engineering text on heat transfer).

If the heating medium is not condensing steam, the assumption of constant T_H is not valid unless the flow rate of heating medium through the jacket is very high, to maintain T_H essentially constant. When this occurs, the differential equation becomes nonlinear and must be solved numerically or graphically.

For the purpose of comparing heating in an agitated vessel with that in a plate heat exchanger, it will be assumed that T_H is constant. This still leaves the problem of finding U. Option 2, literature values, is the most practical for this problem. An overall heat transfer coefficient on the order of 100–150 BTU/hr ft² °F is reasonable for the problem under consideration (19). In the plate heat exchanger, the flow rate was 1,000 gal/hr. A jacketed vessel with a diameter of 5.5 ft and a height of 6.0 ft would be adequate for a 1,000 gal batch. What time would be required to heat the batch from 70°F to 175°F? Assume that the jacket contains condensing steam at 240°F and that mixing is complete so that the tank contents are at a uniform temperature. Assume a heat transfer coefficient of 150 BTU/hr ft² °F.

Equation 2 can be used to calculate A, the heating time. A total of 1,000 gal will occupy 94% of the vessels volume so that the area for heat transfer will be:

$$A = 0.94 \left(\pi DL + \pi \frac{D^2}{4} \right) = 127.4 \text{ ft}^2$$

Then

$$\theta = \frac{WC_p}{UA} \ln \left(\frac{T_H - T_0}{T_H - T_f} \right)$$

and after substitution, $\theta = 1384$ sec.

The difference in residence times is quite significant. One could expect undesirable reactions to occur during this long time period. This is particularly true close to the wall of the vessel where local temperatures would be considerably higher than the bulk average. Obviously, this problem is influenced by vessel size. What would happen if the batch were divided between two vessels? This would require two vessels of 4.25 ft in diameter by 5.0 ft in height to maintain roughly the same proportions. The calculated residence time for a half-size batch is 0.32 hr or 1,160 sec.

For heat-sensitive products such as this, excessive heating is to be avoided.

Plate heat exchangers are not without their problems. Perhaps the most serious is fouling of the heat exchange surface by burn-on of heat-sensitive materials. This reduces the heat transfer coefficient with time and induces a transient effect over a process that is considered to be at steady state (13).

V. WESTERNIZATION OF THE SOYBEAN ~~Bo~~ Inedible oil
~~Denatured:~~

Growth of the soybean industry in the United States was not based on products like tofu. Soybean contains about 18% edible oil. This oil was the "reason for being" for soybean. The first process for recovery of soybean oil was mechanical expression with screw presses. It has been replaced by solvent extraction to increase yields of oil. Mechanical expression leaves 2—3% oil in the press cake under the best of conditions (21). Solvent extraction is capable of reducing oil content to less than 1%. The cost savings in increased yields are such that the greater capital investment and operating cost of a solvent extraction plant is justified. Virtually all soybean oil in the United States is obtained by solvent extraction. Data gathered by the U.S. Department of Agriculture (23) shows a 12% increase in yield for solvent extraction versus mechanical expression. Although the reason for introducing solvent extraction was based on oil recovery, it also provided the technology necessary to develop the protein fraction for human use.

Mechanical expression unavoidably develops relatively high temperatures, with the resultant press cake being unsuitable for human consumption due to the development of off flavors. In addition, the protein in the press cake is denatured and not easily separated from the carbohydrate fraction. Temperature conditions during solvent extraction are mild and controllable so that it is possible to maintain the protein in its "native" state. This opens the door to many opportunities for food use.

A. Solvent Extraction

A typical flow sheet for a hexane solvent extraction plant is shown in Fig. 3. Hexane extraction plants are generally sized to process 3,000 tons/day of soybean.

Cleaned soybeans are cracked with corrugated rolls running at differential speeds. The hulls are effectively loosened in this manner and can be separated by aspiration. The cracked beans, or grits, are tempered prior to flaking. Moisture is raised to 10—11% and temperature to 160—170°F so that the grits are somewhat plastic. Flaking is accomplished with smooth milling rolls similar to

dehull =
hull =

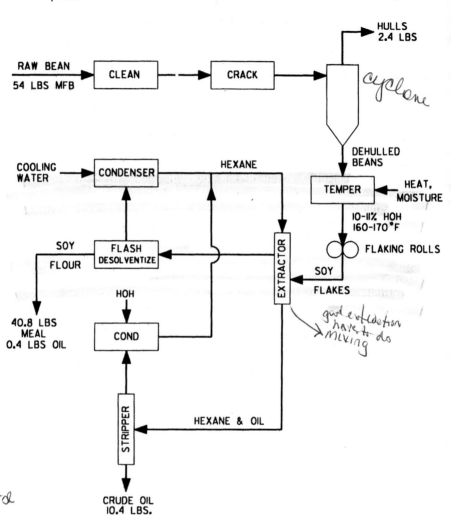

cyclone

good extraction
have to do
mixing

extractor

FIGURE 3   Hexane solvent extraction flow sheet.

those used to manufacture flaked breakfast cereals. Flake thickness
is a critical variable since in commercial extractions the rate is diffusion-limited in the critical region of low oil content. Although
minimum thickness is desirable from a diffusion standpoint, mechanical strength of the flake is also an important consideration for
maintaining good solvent flow through the bed of flakes. It is also

important that flake integrity be maintained to avoid the formation of fines, which would result in material handling problems. Flake thickness is generally held in the range of 0.01–0.016 in. (25).

As with most natural food products, variability and nonideality is generally the rule so that theory can be applied but it must be done with an understanding of all the phenomena that are occurring. Such is the case if one attempts to apply diffusion theory to the extraction of oil from flakes. The complexity of the situation has been described previously (2,7,16) and merits discussion here for it illustrates some of the difficulties encountered in dealing with natural products.

For thin platelets, homogeneously impregnated with oil the unsteady state diffusion equation can be written as

$$E = \frac{8}{\pi^2} \sum_{n=0}^{\infty} \frac{1}{(2n+1)^2} \exp - \left\{ (2n + 1)^2 \left(\frac{\pi}{2} \right)^2 \frac{D\Theta}{R^2} \right\} \qquad (3)$$

where E = fraction of total oil remaining after time Θ (hr); R = half plate thickness (ft); D = diffusion coefficient (ft^2/hr). It is assumed that end effects, negligible area for mass transfer from edges of the platelet, can be ignored.

Except for low values of Θ, this equation has the approximate form as follows (2):

$$E = \frac{8}{\pi^2} \exp - \left\{ \frac{\pi^2 D\Theta}{4R^2} \right\} \qquad (4)$$

Thus, a semi-log plot of E versus Θ should give a straight line of slope D/R^2.

It is interesting to compare this equation with some experimental results. These are shown in Fig. 4.

The experimental curve for peanuts is for uniform slices obtained with a microtome (7). Initially, the curve is nonlinear, until a large proportion of the oil has been removed (75%) in a short time (20 min); beyond this point the model is valid. This has been explained in terms of destruction of cell walls during cutting of the platelets, which in effect liberates a portion of the oil. Subsequent experiments with varying thicknesses showed that the fraction of easily extracted oil decreased as thickness increased. Thus, as fewer cells are disrupted, the nonlinearity is less pronounced.

The curve for cottonseed flakes differs substantially from theory in two ways. First, there is the same nonlinearity in the initial portion as with peanut slices, but it is more pronounced (i.e., more nonlinear). Presumably, this is due to even greater disruption of fat-containing cells during flaking as compared to slicing with a

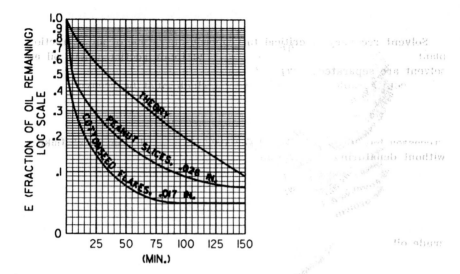

FIGURE 4   Extraction curves comparing theory and experimental data.   (Adapted from Ref. 16, used with permission.)

microtome as in the example with peanuts.   Also note that in the same 20 min almost 90% of the oil has been extracted.   Second, the curve is continuously concave upwards from the time axis.   There are several possible explanations for this observed phenomenon. These include structural heterogeneity leading to two separate diffusion processes; nonuniformity in flake thickness, which would essentially be an artifact; and a change or shift in solubility of the "last" remaining oil.   It is known that the extracted oil from the end of the extraction process is higher in phophatides and other nonglycerides and has lower solubility in the solvent.   A chemical composition change could explain the observed results.

What is the practical significance of these theories?   Regardless of the theory, it is observed in practice that the extraction rate in the range of 0.5–5.0% residual oil is very slow and, in fact, is the controlling rate in extractor design (16).   Since one is most interested in this range for economic reasons, understanding of the physical phenomenon becomes very important.   There is obviously the need for a good model, but it must incorporate the realities of extraction from flaked natural products.

B.   Extractors

Considerable effort has gone into the design of continuous counter current extractors starting with early development in Germany immediately following World War I. Several types of extractors are in use today and are described elsewhere (16).

Solvent recovery is critical to economic operation of an extraction plant. Primary recovery occurs in the stripper section where oil and solvent are separated. Spent flakes coming from the extractor contain a considerable amount of hexane, which must be recovered. It is relatively easy to accomplish this by steam stripping, but in so doing the solubility or dispersibility of the protein in the spent flakes is reduced and its value as a food ingredient is diminished. Processes have been developed for flash desolventizing spent flakes without denaturing the protein (25).

VI. PROTEIN FRACTION

At this juncture, the soybean has been split into two fractions; a crude oil, which will ultimately yield cooking oils, shortenings, salad dressings, lecithin, and margarine, and a protein fraction, which will become concentrates, isolates, and texturized meat extenders and analogs. We will follow the path of the protein fraction, for it is abundant in examples of engineering applications that apply to a broad spectrum of the food processing industry. Those with a specific interest in oil refining will find a good summary with extensive bibliography elsewhere (21).

A. Soy Flour

The most common form of defatted soybean is called "soy flour," which is a misnomer since it is not a flour in the same sense as wheat flour. It does, however, resemble flour in appearance and flow properties. It is ground fine enough so that 100% will pass a number 100 U.S. standard screen, and 60% will pass a 300 mesh screen. Soy flour of this particle size will fluidize readily in pneumatic systems. Although particle size is important for subsequent operations involving mass transfer, the condition of the protein is of prime importance. The composition of soy flour varies from year to year and among varieties. The composition of an average product would be as follows:

Protein	53%
Carbohydrate	30
Fat	1
Fiber	3
Ash	6
Moisture	7

The condition of the protein is specified in terms of either a nitrogen solubility index (NSI) or a protein dispersibility index (PDI).

A value of 100% would indicate a completely native protein. High NSI/PDI soy flour is characterized by protein that is undenatured with high solubility of the protein in water. Many food processing applications require high NSI soy flour. In its normal form, high NSI soy flour has several deficiencies:

Taste and odor become objectionable as soon as water is added.
Nutritional inhibitors (trypsin inhibitors) are present and active.
Texture and appearance are not very appealing.

Fortunately, these deficiencies can all be corrected to a point where organoleptically acceptable, nutritious products can be manufactured from soy flour. In so doing, value will be added to a commodity to make it more attractive and useful to the consumer. In addition, a value-added product can support price differentiation and higher profit margins than a commodity.

To illustrate, suppose that a 2-lb bag of wheat flour costs $0.75. If you add shortening, leavening, and a few other minor ingredients to that flour so that it now functions as a convenient, all-purpose baking mix, you have added "value." That 2 lb of flour with some added cost for other ingredients will sell for $1.35.

How can value be added to soy flour? There are several paths for improving soy flour, as depicted in Fig. 5.

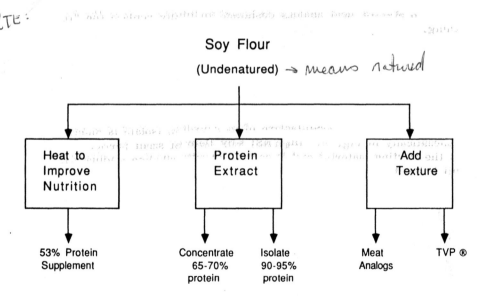

FIGURE 5   Value-added paths for soy flour.

Simply heating soy flour to inactivate trypsin inhibitor permits its use in products such as cereals, breads, cookies, and the like as a nutritional supplement. The protein is denatured but still capable of entering into Maillard reactions with reducing sugars so that usage level is governed to some extent by browning. Flavor is another issue in bland products. The addition of soy flour, with its high lysine content, to wheat-based foods not only raises the protein level but has a synergistic effect by supplying an amino acid that is limiting in wheat. Levels are limited in bread where the addition of high levels of soy flour interferes with formation of the wheat gluten matrix with a resultant decrease in loaf volume. Usage is generally in the 2–5% level in breads. Even at such low levels, the soy flour provides significant quantities of protein. Addition of 2% soy flour will increase the protein content of the fortified wheat by about 7%.

The protein content can be increased, and flavor and odor improved by separating the proteins from the carbohydrates of soy flour. Two types of products are manufactured: isolates and concentrates. Isolates contain 90–95% protein, whereas concentrates are at 65–70% protein. These products find application in breakfast cereals, emulsion meat products, infant formulas, and milk replacers, and as meat extenders, to name a few.

The ultimate refinement of soy flour is the addition of texture in such a manner as to simulate meat. The products that have been manufactured range from utilitarian TVP for extending comminuted meats to elegant meat analogs designed to totally replace the "real" thing.

The level of engineering sophistication required to produce concentrates, isolates, TVP, and meat analogs is often surprising to those not familiar with the food processing industry.

B. Soy Protein Isolate

The process for the manufacture of soy protein isolate is shown schematically in Fig. 6. High NSI soy flour of small particle size is the starting material and is contacted with alkaline medium with pH below 9. Conditions are selected to get optimum yields. The pH is maintained below 9 to prevent hydrolysis of the protein. The isolate process is effective because of the solubility characteristics of soy protein. The solubility curve for soy protein as a function of pH is shown in Fig. 7. Note that the solubility at pH 9 is approximately 90%. When the protein has been dissolved, the slurry is centrifuged to separate or "isolate" the protein from the carbohydrate. The liquor, containing the protein fraction, is next acidified to a pH near the isoelectric point of 4.3 with hydrochloric, sulfuric, phosphoric, or acetic acid, which causes the protein to

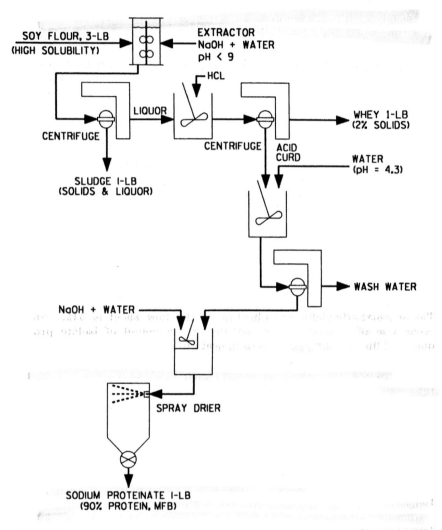

FIGURE 6 Soy protein isolate.

precipitate. Note that the protein has a solubility of 10% at this pH.
The acid curd is centrifuged to separate it from the whey, washed
(pH 4.3), and once again centrifuged. The most common form of
isolate, the sodium proteinate, is formed by neutralizing the curd
with sodium hydroxide followed by spray drying.

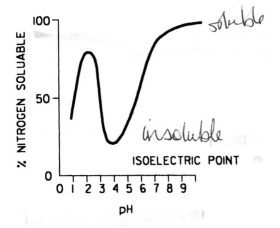

FIGURE 7   Solubility of soy protein versus pH.

C. Economics

The approximate yield as indicated on the flow sheet is 33%. Another way of stating this is that for every pound of isolate produced, 2 lb of solid waste are generated. The sludge stream contains approximately 15% solids, both suspended and dissolved. The
protein content of the solids can reach 40%. The protein is largely
dissolved in the sludge liquor, and the solids consist primarily of a
complex blend of carbohydrates. The whey stream is very dilute
with a solid content ranging from 1% to 3% depending on the solids/
liquids ratio used in the extractor. These waste streams are a serious problem since they have high biological oxygen demand (BOD)
and, in the case of the sludge, high suspended solids also. There
are two issues to resolve:

1. Is there a cost savings opportunity? Is it possible to recover
    some of the lost protein and thereby increase yield?
2. There is considerable concern over and regulation of discharge
    of food processing waste into municipal sewage systems. There
    are usually three options to consider.
    a. Treat the waste in your own plant. The choices are to
        recover the waste and reuse or build a waste treatment
        plant.
    b. Pay the toll charge for discharges into the municipal sys-
        tem.
    c. Go out of business.

A waste treatment problem should be attacked in stages. First, seek opportunities to reduce waste and recover/recycle as much as possible. Look for the cost savings opportunities. Second, when these are exhausted, deal with the waste disposal problem.

D. Waste Recovery *Waste becomes product*

There is an opportunity to recover a substantial portion of the valuable protein in the "sludge" stream. The sludge is comprised of two parts: suspended solids, which are primarily carbohydrate and cellulosic polysaccharides, and the liquor, which contains dissolved protein in the same concentration as the clear effluent from the first centrifuge. The solution is the same as recovering liquor from the filter cake in a filtration operation. If the sludge is mixed with an excess of fresh water, a certain fraction of the dissolved protein can be "washed out" and recovered by centrifugation. The extent of recovery depends on the amount of water used to "wash" the sludge. There is a practical limit since dilution becomes a factor in dictating capital costs for centrifuges and holding tanks. The following problem illustrates the principle.

It is proposed to install a washing tank, centrifuge, and necessary pumps and piping to recover protein from the sludge liquor. Current feed rates are given in Fig. 8. The system is to be sized such that the sludge liquor can be diluted 2:1. All the protein in the sludge is dissolved in the liquid phase. Soy flour costs 12¢/lb; the corporate income tax rate is 50%; and the equipment has an economic life of 10 years. The plant operates 300 days/year, with an efficiency of 83% (20 hr/day). The plant proposes to maintain the production rate of isolate at 381 lb protein/hr, and, accordingly, the recovered protein will be used to reduce the soy flour requirement. We will examine the following questions:

1. What is the yearly cost savings with the new equipment?
2. What is the ROI for this proposal?

Overall Material Balance on Protein

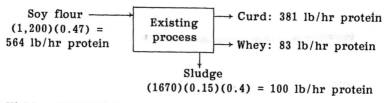

Soy flour ──────→ | Existing process | → Curd: 381 lb/hr protein
(1,200)(0.47) =
564 lb/hr protein → Whey: 83 lb/hr protein

Sludge
(1670)(0.15)(0.4) = 100 lb/hr protein

Yield = (381 lb/1,200)100 = 31.75%

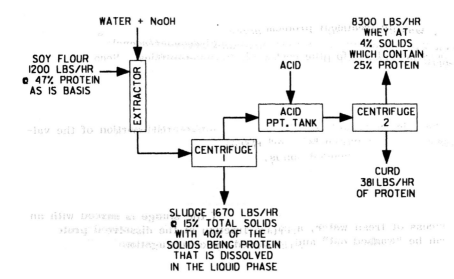

FIGURE 8   Flow sheet for existing extraction plant.

Basis: 100 lb/hr is dissolved in sludge liquor. This is diluted 2:1 with water so that two-thirds are recoverable by subsequent centrifugation: 2/3(100) = 67 lb/hr.

In computing a revised yield, note that some of the recovered protein is lost to the whey stream during acid precipitation. Thus, the revised yield is as follows:

$$\text{Recovery} = 381 + 67 \frac{381}{(83 + 381)} = 436 \text{ lb/hr protein}$$

$$\text{Yield} = \frac{436}{1,200} \times 100\% = 36.33\%$$

To produce 381 lb/hr of protein, the soy flour requirement is 381/0.3633 = 1,048.7 lb/hr, and savings = (1,200 - 1048.7)(0.12)-(20)(300) = \$108,936/year.

The capital cost has been estimated by first developing a detailed flow sheet as shown in Fig. 9.

The contingency (Table 2) is included to cover unforeseen events such as price changes between first estimate and firm bid, variances

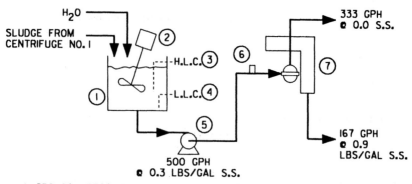

I. 250 GAL. STAINLESS STL.   5. 600 GPH 100 PSI PUMP W/15 HP MOTOR

I. 250 GAL. STAINLESS STL.   5. 600 GPH 100 PSI PUMP W/15 HP MOTOR
2. 3 HP AGITATOR   6. PRESSURE SWITCH
3. HIGH LEVEL CONTROLLER   7. CENTRIFUGE STAINLESS STL. CONST.
4. LOW LEVEL CONTROLLER   S.S. = SUSPENDED SOLIDS

FIGURE 9   Specification flow sheet for washing equipment.

between bid and estimate for equipment installation, code variances, design errors, and the like. Contingency is not intended to cover changes in scope, such as the need for a second centrifuge if a single unit proved to be inadequate. The amount of the contingency is inversely related to the quality or accuracy of the capital estimate. If a preliminary engineering design is provided to contractors for bidding and equipment quotes are firm, the contingency may be in the range of 5–10%. If installation costs are obtained by applying multipliers to telephone estimates of equipment costs, contingency would be in the 20–30% range. It is always preferred, if time permits, to develop a preliminary engineering design prior to the cost estimate. As is often the case, time is at a premium and estimates are developed without benefit of engineering design so that contingency must be high to cover the unknowns.

To simplify the ROI calculation, assume that the cash flow is constant from year to year. The resulting ROI will not be sufficient to satisfy financial analysts scrutinizing a capital proposal, but it is close enough to indicate if the proposal has sufficient merit to be seriously considered.

The pertinent formula is Eq. 5 (of Chapter 3) in conjunction with the present value factor table (Table 3 of Chapter 3), with

I   = \$172,000
N   = 10
BTE = \$108,936

TABLE 2 Cost Estimate

Item	Installed cost
1. 250 gallon stainless steel tank	$10,000
2. 3 HP agitator	1,000
3. High level control	700
4. Low level control	700
5. 600 GPH stainless steel pump, 100 psi	8,000
6. Pressure control switch	700
7. Centrifuge stainless steel const.	85,000
8. Piping, stainless steel	5,500
9. Electrical	
3 motors	12,000
3 controls	1,500
Equipment, total	125,100
Engineering (15%)	18,500
Subtotal	143,600
Contingency (20%)	28,700
	172,300

Source: P. K. Sherman, Pillsbury Co. (1985).

Tax $= 1/2(108,936 - 17,200) = \$45,868$
CF $= 108,936 - 45,868 = \$63,068$

Then,

$$PF = CF \sum \left(\frac{1}{1+i} \right)^n$$

$$172,000 = 63,068 \sum \left(\frac{1}{1+i} \right)^n$$

and

$$\sum \left(\frac{1}{1+i} \right)^n = 2.73$$

From Table 3 (of Chapter 3) for 10 years,

$$\sum \left(\frac{1}{1 + i}\right)^n = 3.09 \text{ for ROI} = 30\%$$

$$\sum \left(\frac{1}{1 + i}\right)^n = 2.38 \text{ for ROI} = 40\%$$

Interpolating, the ROI is approximately 35%.

This option appears to have considerable merit and would be developed to a higher level of accuracy before an investment decision would be made. Recognize that this does not resolve the waste disposal problem.

VII. WASTE TREATMENT

The cost of waste disposal is highly variable depending on location and type of waste. To gain some insight into the complexity of the problem, consider the streams being discharged by the soy isolate plant (Fig. 8). The easiest solution to the waste problem is to discharge to the municipal sewage treatment plant. This generally requires little or no capital. Charges are assessed on the basis of BOD, suspended solids, and volume of discharge. It is helpful to have some measure of the impact of a plant such as this on a sewage treatment system. Typically, six people generate 1 lb BOD_5/day. The plant in this example is discharging 11,650 lb/day of solids that will have a BOD_5 equivalent of approximately 11,650 lb/day.

This is the equivalent of 69,900 additional people added to the sewage treatment system. Obviously, cities, and especially smaller ones, have to anticipate industrial waste loads when planning municipal treatment plants. As an order of magnitude estimate, if the sewage charge were on the order of 3¢/lb BOD and 3¢/lb suspended solids, this plant would incur sewage costs of some $440/day or $132,000/year. This is a significant operating cost. If we allocated sewage cost back to the product, it would add almost 6¢/lb of product. Applying a 25% margin to this translates to a price increase of 8¢/lb for the consumer of the product. Generally, that consumer is another manufacturer who uses the isolate as an ingredient in a consumer product. Typically, the second manufacturer's margin would be on the order of 35% so that the effect of the sewage treatment is now 12.3¢ per pound of isolate to the ultimate consumer—you.

Competition in the market place is such that those striving to maintain low-cost producer status will search for economic alternatives to avoid discharge in municipal sewage systems. It is a simple case of economics, and it does clearly point out why strong pollution laws and strict enforcement are necessary to prevent abuse of the waterways. There are two problems to deal with here. The sludge is fairly concentrated at 15% solids, and with 40% protein content has the potential of being a good animal feed, especially for ruminants. The whey is very dilute by contrast, contains only 4% solids, and also contains a high percentage of trypsin inhibitor as well as soluble oligosaccharides.

A. Sludge Conversion to Animal Feed

Consider first a proposal to convert the sludge to animal feed by drying in a flash dryer and packaging the dry material in 100-lb bags. It is not known how much a flash drying system will cost, and you are interested in determining how much could be spent on capital to support the proposal. A reasonable return on investment would be 20% after taxes. The corporate tax rate is 50% and economic life of the investment would be 10 years. Flow rates and other parameters are those given in Fig. 8.

The product will be a granular, free-flowing solid at 8% moisture with a protein content of 37% on an as-is basis. It will be packed in 100-lb bags, which cost $0.35 each. Actual product value would have to be assessed through feeding studies. For estimation purposes, it will be assumed that the protein quality is the same as soy flour, and its value will be factored from the selling price of 44% feed grade soy meal at $180/ton: selling price = 37/44 ($180) = $151/ton.

1. Labor/Operations

The plant will be operated on a two-shift basis with one person per shift. Average wage is $20,000/year. Total wages are $40,000/year. Operating costs including incidental utilities, insurance, and space heating are $15,000/year.

2. Drying Costs

Total sludge = (1670)(20)300 = 10,020,000 lb/year

$$\text{Solids @ 8\% moisture} = \frac{(0.15)(10,020,000)}{0.92} = 1,633,700 \text{ lb/year}$$

Moisture = $(10.020 - 1.634)10^6$ lb/year = 8.386×10^6 lb/year to be removed

BTU requirement = $(8.386 \times 10^6)(212 - 70)(1) + 1,150 = 10.8347 \times 10^9$ BTU/year

A dryer efficiency of 50% is assumed for the flash dryer. This is understated but is balanced by the selling price, which is probably overstated.

BTU for dryer = $2 \times 10.8347 \times 10^9 = 21.67 \times 10^9$ BTU/year

The dryer will use natural gas as a heat source. Gas heating value is 1,000 BTU/ft^3 or 10^6 BTU/MCF (MCF = 1,000 ft^3).

Rate = \$4.00/MCF (U.S. average)

Energy cost = $\dfrac{2.167 \times 10^{10}}{10^6}$ (\$4.00) = \$86,700/year

3. Bags

Production requires 16,500 bags, with 1% shrink on bags. Cost = 16,500(\$0.35) = \$5,775/yr.

4. Year Operating Cost

The yearly operating costs are such:

Gas	\$ 86,700
Labor	\$ 40,000
Bags	\$ 5,775
Miscellaneous	\$ 15,000
Total	\$147,475

5. Sales

The yearly sales revenue is as follows: (16,338 cwt)(\$7.55/cwt) = \$117,184. The 1% penalty is to offset distribution losses (trucking to warehouses).

6. Savings on Sewage Charge

The sewage charge for the sludge stream, based on 3¢/lb BOD and 3¢/lb suspended solids would be as follows:

Sewage charge = (1670)(0.15)(0.6)(\$0.03) + (1670)(0.15)(\$0.03) = \$12.02/hr
Yearly charge = (12.02)(20)(300) = \$72,144

This cost savings can be added to the income from the dry sludge with a total savings of \$117,814 + \$72,144 = \$189,958.

7. ROI/Capital Investment

To calculate the capital that yields an ROI of 20% (after tax) refer to Eq. 5 (of Chapter 3) and Table 3 (of Chapter 3).

BTDE = 189,958 − 147,475 = 42,483
DEP = X/10, where X = capital cost

Tax = $1/2$ (42,483 - X/10)
CF = 42,483 - $1/2$ (42,483 - X/10) = 21,241 + X/20

From Table 3 (of Chapter 3), the present value factor, $\Sigma(1/1 + i)^j$, for n = 10 and i = 20 is 4.192. Then:

PV = X = CF $\Sigma(1/1 + i)^n$
X = (21,241 + X/20) (4.192)
Solving for X, X = \$112,714

It is highly unlikely that \$112,714 is enough capital to install a flash dryer and associated equipment. The factor that causes the proposition to fail is the cost of energy for drying. For example, if energy costs could be reduced by 50%, to a level of \$43,000/year, the proposal would sustain a capital investment of \$230,000 at a 20% ROI and \$275,000 if one would accept a 16% ROI. This latter figure might be very attractive if a nagging problem could be eliminated.

What is the potential for a 50% reduction in energy usage? Two steps can be taken. First, the 50% efficiency of the flash dryer is certainly conservative. Flash dryers with efficiencies in the 75–80% range are available. A dryer with 75% efficiency represents a 33% reduction in energy costs. Second, mechanical dewatering of the sludge such as with screw press or filtration should be investigated. These will require some capital and experimental trials.

B. Economics of Waste Treatment

Pollution abatement projects are generally not very attractive financially. If they were, there would be no need for laws regulating effluent discharge. An extra effort is often required to find innovative techniques to improve the economics.

The whey stream is a particularly difficult problem for several reasons. Being very dilute, it must be concentrated before drying is even considered. Technically, the whey can be concentrated by techniques such as ultrafiltration/reverse osmosis (9).

This is done with whey from milk with the by-product being a very high quality protein with excellent functional properties. However, with soy isolate whey, trypsin inhibitor would most likely be included in the protein concentrate and would be a concern in food applications requiring a functional protein. No economical solutions have been found to date (6).

VIII. SOY CONCENTRATES

Concentrates contain 65–70% protein. The concept for producing a concentrate is the reverse of that for an isolate. For an isolate, the protein fraction is solubilized and separated from the insoluble carbohydrate. Concentrates are produced by dissolving the carbohydrate fraction, leaving the protein intact. There are three methods for manufacturing concentrates:

1. *Isoelectric wash*: Carbohydrate is leached from the protein with aqueous acid in the isoelectric range (near pH 4.3) where the protein has minimum solubility.
2. *A 20—80% aqueous alcohol leaching*: Proteins are insoluble in this range, whereas carbohydrates are soluble.
3. *Heat*: Heat denatures the protein to insolubilize followed by water leaching to remove carbohydrates.

The first two methods result in high NSI concentrates; the third does not. In processes 1 and 3, the leaching water can be discarded consistent with pollution regulations. In the case of aqueous alcohol leaching, the solvent cannot be discarded for economic reasons. Yields are in the 60—70% range.

IX. PROTEIN TEXTURIZATION

The prospect of simulating meat, either as an analog or simply as an extender, from vegetable protein has received considerable attention from the food processing industry. There have been major research and development efforts in the areas of meat analogs and meat extenders with untold millions of dollars spent to come up with the winning product or process.

One can gain a considerable insight into a technology from the patent literature. In the case of protein texturization, the patent literature is abundant and quite interesting. The list of companies that have obtained patents related to either meat analogs or meat extenders reads like a who's who in the U.S. food industry and includes Archer Daniels Midland, Central Soya, General Foods, General Mills, Griffith Labs, Lever Brothers, Miles Laboratories, Nestle', Pillsbury, Proctor and Gamble, Quaker Oats, Ralston Purina, Staley, Swift, and Worthington Foods.

The idea of producing high-quality meat analogs and extenders from inexpensive vegetable proteins not only captures the imagination of the researcher but offers some significant financial rewards to those who can succeed.

A. Spun Protein Fibers/Meat Analogs

The most elegant and probably most expensive way to impart meat-like texture to soy protein is by spinning, as in the rayon process.

to form fibers. These fibers can then be manipulated and combined with other ingredients to impart a texture similar to meats like ham, chicken, and beef. Figure 10 illustrates a continuous process for spun protein fibers.

The first part of the process is very similar to the isolate process. The soy flour is high NSI, and SO_2 is used to form the curd

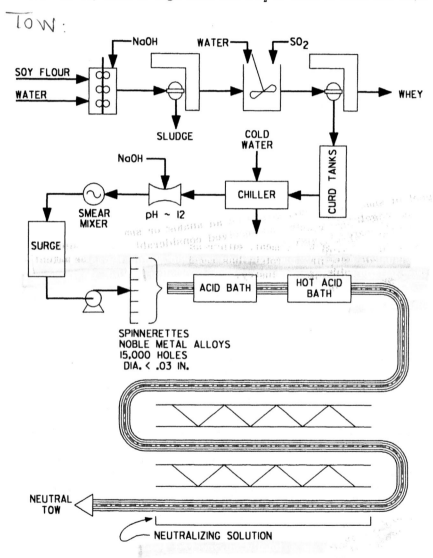

FIGURE 10 Flowsheet for manufacture of spun protein fibers.

instead of HCl. The use of SO_2 is related to improved textural
characteristics and flavor in the resulting fiber (24). The acid
curd can be washed to remove solubles, after which it is held in
a curd tank.

The acid curd is mixed with NaOH in a specially designed mixer
which is shown in Fig. 11.

The mixing device is a screw pump with mixing, transition, and
metering sections. The curd has been converted to a clear solution
at pH approximately 12, which is referred to as spinning dope. The
dope is held in a surge tank prior to being metered through the
spinnerettes, which are constructed of noble metal alloys and contain 15,000 holes with diameters of 0.003 in. The holding time between the screw pump and the spinnerettes is less than 5 min (24).
It is important to maintain a short holding period to minimize hydrolysis and subsequent rheological problems. The fibers emerge
from the spinnerettes into an acid bath containing acetic acid. The
pH is such that the protein becomes insoluble at this point. The
fibers are carried over rolls, are stretched, and finally enter a second acid bath maintained at a temperature high enough to denature
the protein and permanently set its structure. Neutralization of
the acid-impregnated fiber tow is a crucial step in the process in
that inadequate removal of the acid will be detrimental to flavor
quality of the finished meat analog.

Visualize a tightly packed fiber tow containing thousands of individual fibers, surrounded by an acid medium. How do you penetrate this fibrous mass to leach out the acid and do so in a continuous process? A mechanical device for continuous neutralization of a
fiber tow is shown in Fig. 12.

The principle employed is the use of a vibratory system capable
of inducing horizontal and vertical forces on the tow sufficient to
effectively separate the fibers and permit leaching to take place.
The horizontal component of vibration also serves to convey the
tow through the device. Flow of neutralizing solution is in a generally countercurrent direction.

The neutralized fiber tow is a half-product at this stage. It can
be sold as such to other manufacturers, or it can be further processed to produce a meat analog. Figure 13 shows a process for
meat analogs such as chicken and ham.

The fiber tow is cut into pieces 1/2–1 in. in length in a cutting device. The concept of forming a meat analog essentially consists of setting the protein fibers in a heat-coagulable protein gel
and adding the proper flavors, fats, and other minor ingredients
to impart a meatlike texture, flavor, odor, and appearance to the
final product. It is a totally engineered food. The flow sheet

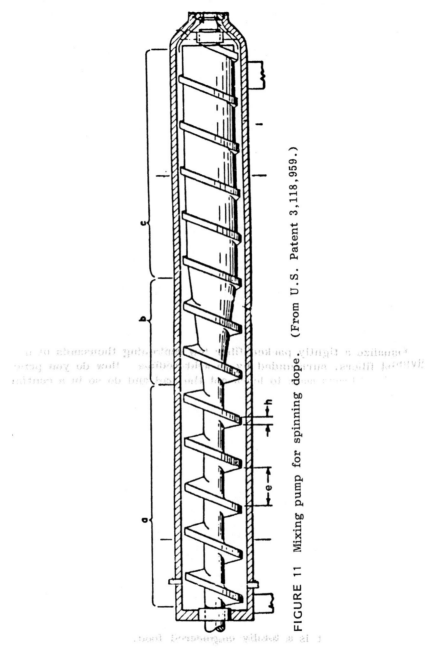

FIGURE 11 Mixing pump for spinning dope. (From U.S. Patent 3,118,959.)

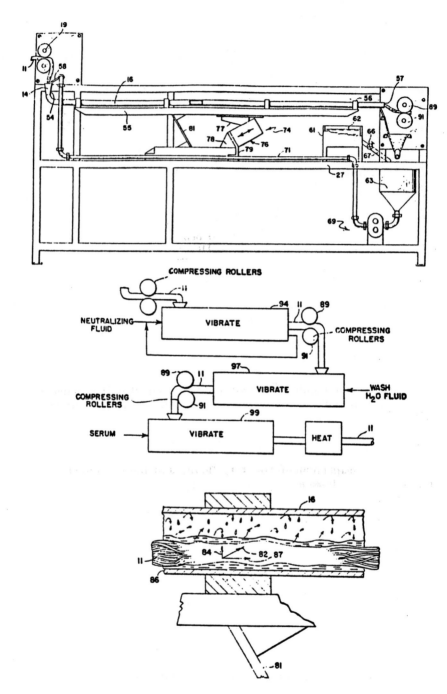

FIGURE 12 Fiber tow neutralizing device. (From U.S. Patent 3,269,841.)

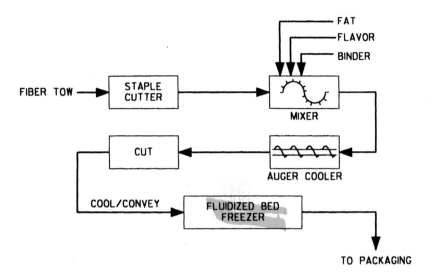

FIGURE 13   Meat analog process.

(Fig. 13) depicts two steps in the forming process: a mixing step to combine fat, flavor, and binders with the protein fibers, and a cooking step to heat-set the mass into meatlike chunks. The process for accomplishing this has been patented (17,18). The specific machinery involved in the mixing and cooking steps is detailed in Fig. 14.

A typical composition of the fat, flour, and binder added in the mixing step is as follows (18):

Water	50%
Fat/oil	20%
Filler cereal (flour, grits)	10%
Egg albumin	10%
Cellulose	4%
Sugar	2.5%
Flavor	3.5%

The ratio is about 55% fat, flavor, and binder solution to 45% cut fiber (wet basis). The egg albumin acts as the protein binder or "food glue." It also provides a good blend of amino acids that synergistically supplement the amino acids of the soy fiber. The

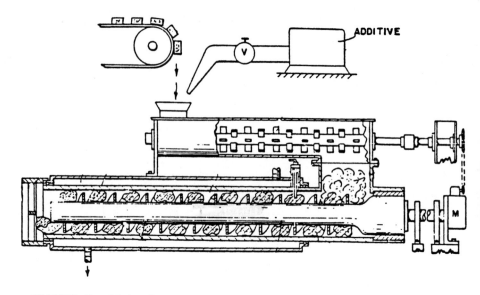

FIGURE 14 Mixing/cooking device for meat analogs. (From U.S.
Patent 3,558,324.)

egg albumin is more than a minor ingredient. Recognizing that the
spun fiber is about 60% water, the soy protein represents about 18%
of the finished product formula and egg albumin represents 5.5%.
It is essential that meat analogs have the same nutritional value of
the foods they are intended to replace. Soy fiber, like soy isolate,
would have a PER of about 1.3 (Table 6 of Chapter 4). The addition of significant quantities of egg albumin will raise the PER to
comparable levels of the meat being replaced (3). Although this
is a specific formula, a more general formulation has appeared in
the literature (22), which indicates a similar proportion of protein
fiber to binder.

   A general composition of meat analogs on a dry basis follows
(22):

Spun fiber, 40%
Protein binder, 10%
Fat, 20%
Flavors and colors, supplemental nutrients, 30%

Although the flavor appears at a low level in the finished product, it is a very critical ingredient. The attempt to develop a flavor system for food products is a substantial undertaking and will usually involve several companies (flavor houses). The success of the product depends to a large extent on the accuracy of the flavor match.

The auger cooker of Fig. 14 is designed so the product is compressed to remove entrapped air. The auger is steam jacketed so that heat can be applied to set the egg albumin binder by denaturation. A temperature range of 140–160°F is adequate to set the albumin.

As the slabs of meat analog are extruded from the auger cooker, they are sliced or diced into convenient size chunks.

The final processing step is freezing. The marketing concept for a diced meat analog implies multiple use from an individual package. This imposes a processing constraint on the plant, since the frozen chunks must be free flowing and individual in the frozen state to allow multiple use without thawing and refreezing on the part of the consumer. To obtain individually frozen pieces requires specialized freezer design. Freezers that are designed to perform this task are air blast freezers where the cold air is directed through a fluidized bed of particulates moving normal to the flow of air (Fig. 15).

Fluidized bed freezers will have good heat transfer coefficients because of high lineal air velocities through the bed. Fluidization

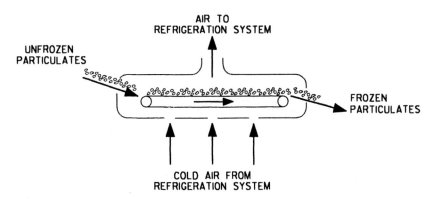

FIGURE 15   Schematic of fluidized bed freezer.

should prevent clumping, and it aids in achieving uniformity through
the bed. There is a limitation on the particulate size that can be
fluidized. Although it is possible to fluidize peas, broccoli will not
fluidize. However, even if fluidization does not occur, the cold air
flow through the bed rather than across is a good design. Piece
individuality may be sacrificed if fluidization does not occur, and it
may be necessary to devise a clump breaker to separate the particulates after they are frozen.

X. FREEZING

Freezing of foods for preservation has increased dramatically in recent years and continues to grow. Walk through the frozen food
aisle of any supermarket and witness the multitude of products to
entice the consumer. From a practical engineering standpoint, the
design of food freezing systems requires two fundamental pieces of
engineering information:

1.   What is the enthalpy change that occurs during freezing? This
     will determine the refrigeration requirement.
2.   How long will it take to freeze the product? With this informa-
     tion, a freezer can be properly sized to provide the required
     production rate.

The second question is generally the more difficult to answer and
most often the one that requires a quick answer.
    Modeling of the heat transfer process during freezing is compli-
cated by the phase change that occurs and the movement of a freezing front from the exterior surface to the interior. This moving
boundary problem has been solved by Neumann (4). The solution
assumes that the surface exposed to the cooling medium is immediately brought to the temperature of that medium. Discussion of
the model and example problems can be found in engineering texts
(5,12). The Neumann solution may be useful in some situations
such as immersion cryogenic freezers. In blast freezers and plate
freezers, the boundary conditions are not valid since a heat transfer coefficient has been introduced. From a practical standpoint,
the engineer is seeking a reasonable approximation as a design
basis. Recognizing the many inaccuracies that are factored into
the problem (shape factors, variability and reliability of thermal
conductivity and specific heat data, end effects, variability in bed
depth, air flow, heterogeneity, and the nonisotropic nature of foods,
etc.), it would seem that the accuracy promised by very elegant
models is not warranted in view of the quality of the data being
used in the models.

The most straightforward approach to calculating freezing times is that of Plank (20).

Several simplifying assumptions are made:

1. Assume constant thermal conductivity (taken as frozen state) throughout.
2. Ignore sensible heat effects, and consider only the phase change at the freezing temperature, t_f.
3. Assume that the entire body is at a uniform t_f just prior to freezing.

For the one-dimensional case of the slab, Fig. 16 gives pertinent dimensions. Writing the heat-conduction equation for the frozen portion of the slab (cross-hatched in Fig. 16),

$$Q = a(t_s - t_f) \ k/x \qquad (5)$$

where

Q = heat flux at surface
k = thermal conductivity
a = surface area
t_s = surface temperature

The heat flux from the surface to the cooling medium can be written as follows:

$$Q = ha \ (t_\infty - t_s) \qquad (6)$$

where h = heat transfer coefficient.

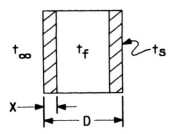

FIGURE 16   Freezing slab.

If t_s is eliminated between Eqs. (5) and (6), the expression for Q is

$$Q = a \; \frac{(t_\infty - t_f)}{\left(\dfrac{1}{h} + \dfrac{x}{k} \right)} \tag{7}$$

The rate of heat generation at the freezing point Q can be expressed in terms of the velocity of the front as,

$$Q = a\lambda\rho \; \frac{dx}{dt} \tag{8}$$

where

dx/dt = velocity of freezing front
λ = latent heat of fusion
ρ = density

Combining Eqs. (7) and (8),

$$\frac{(t_\infty - t_f)}{(1/h + x/k)} = \lambda\rho \; \frac{dx}{dt}$$

and integrating between appropriate limits,

$$\frac{t_\infty - t_f}{\lambda\rho} \int_0^{\Theta_f} dt = - \int_0^{d/2} \left(\frac{1}{h} + \frac{x}{k} \right) dx$$

which after simplifying gives

$$\Theta_f = \frac{\lambda\rho}{(t_f - t_\infty)} \left[\frac{d}{2h} + \frac{d^2}{8k} \right] \tag{9}$$

where Θ_f = time to achieve complete freezing temperature t_f.
Equation (9) is usually written in the generalized form as,

$$\Theta_f = \frac{\lambda\rho}{(t_f - t_\infty)} \left[\frac{Pd}{h} + \frac{Rd^2}{k} \right] \tag{10}$$

where

P = 1/2, R = 1/8 for an infinite slab
P = 1/6, R = 1/24 for a sphere of diameter d
P = 1/4, R = 1/16 for an infinite cylinder of diameter d

For a parallelopiped geometry, the constants P and R are obtained from the chart shown in Fig. 17.
If heat capacity data are available for the material in question, it is desirable to use a corrected form of Plank's equation to account for sensible heat effects above and below the freezing temperature. One such modification, due to Nagaoka (15) is,

$$\Theta_f = \frac{\Delta H' \rho}{t_f - t_\infty} \left[\frac{Pd}{h} + \frac{Rd^2}{k} \right] \tag{11}$$

and,

$$\Delta H' = [1 + 0.00445(t_i - t_f)] \, [Cp_u \, (t_i - t_f) + \lambda + Cp_f \, (t_f - t)]$$

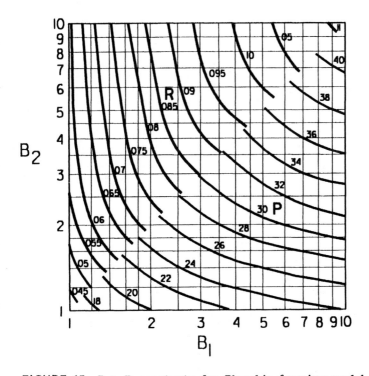

FIGURE 17 P & R constants for Planck's freezing model.

where

t_i = initial temperature of object
t = final temperature of frozen object
Cp_u = heat capacity of unfrozen object
Cp_f = heat capacity of frozen object

To utilize Eq. 11 or 10, one must know the freezing point and the heat of fusion. If the heat of fusion is not available, it can be estimated by calculating a heat of fusion based on the amount of wa-- ter in the products: $\lambda = f\lambda_0$ where λ_0 = heat of fusing ice at t_f (BTU/lb), and f = fraction of water in the product.
Determination of the freezing point can be accomplished several ways:

1. Calculate it based on freezing point depression due to solutes in the product (12).
2. Use available enthalpy/composition charts and construct chart of mass fraction ice versus temperature to find break point (5).
3. Measure experimentally.

All the methods have drawbacks. Method 1 requires knowledge of the composition of the product. Method 2 assumes enthalpy/composition charts are available, which they generally are not. Method 3 requires careful experimentation to determine freezing point.

Example

Returning to the fluidized bed freezer and the meat analog chunks, let's apply the modified Plank's equation to a frequently encountered problem.
Assume that the plant is producing meat analog in 1/2-in. cube form and that the following conditions prevail:

Temperature entering freezer	= 70°F
Temperature exiting freezer	= 0°F
Freezing temperature of analog	= 26°F
Temperature of air blast	= -30°F
% moisture in product	= 70%
Heat capacity unfrozen analog	= 0.84 BTU/lb °F
Heat capacity frozen analog	= 0.49 BTU/lb °F
Thermal conductivity	= 0.64 BTU/hr ft °F
Heat of fusion ice @26°F	= 144 BTU/lb
Density	= 65 lb/ft³
Average residence time	= 5 min

The freezer is operating at the maximum capacity air velocity. Marketing has expressed interest in changing the piece geometry from a 1/2-in. cube to a 1/2-in. × 1/4-in. × 1-in. piece. If bed depth were maintained at current levels and other freezer variables were held constant, how would this affect the rated output of the freezer? What is the freezing time for the new piece geometry?

Solution

The heat transfer coefficient is not given but can be calculated from the data given for the 1/2-in. cube from Eq. 11:

$$\Delta H' = [1 + 0.00445 (70 - 26)]$$

$$[0.84 (70 - 26) + 0.7 (144) + 0.49 (26 - 0)] = 180$$

$$\beta_1 = \beta_2 = 1, \text{ from Fig. 17, R} = 0.43, P = 0.17$$

$$\Theta_f = 0.083 \text{ hr} = \frac{(180)(65)}{26-(-30)} \left[\frac{(1/24)(0.17)}{h} + \frac{(1/24^2)(0.043)}{0.64} \right]$$

then, solving for h, h = 25 BTU/hr ft^2 °F

For the proposed piece, the volume is the same as the 1/2-in. cube so the output from the auger cooker will be the same. For this geometry, $\beta_1 = 2$, $\beta_2 = 4$ and R = 0.08, P = 0.28; $\Delta H'$ is unchanged.

Then,

$$\Theta_f = \frac{(180)(65)}{26-(-30)} \left[\frac{(1/48)(0.28)}{25} + \frac{[(1,48)^2] \, 0.08}{0.64} \right]$$

$$= 0.06 \text{ hr} = 3.6 \text{ min}$$

The residence time to achieve 0°F is less than it was for the cube. There is actually some upside potential with the new geometry. This is no surprise since the critical dimension for conduction was cut in half. Several things would have to be verified with the new geometry before a throughput increase could be implemented:

Will the new piece size geometry be fluidized?
With a thin (1/4-in.) dimension, will the product fracture in the freezer?
Is there excess refrigeration capacity to support a potential increase in mass throughput?
Can the auger cooker run faster?
Is the packaging line capable of running faster?

Some of these can only be answered experimentally in a plant trial or pilot plant run.

XI. PROLOGUE FOR MEAT ANALOGS

The market for soy-based meat analogs has not developed to the extent necessary to support any significant business venture. Hindsight is usually 20/20, and it is not difficult to discern the principal reasons for this failure.

1. The capital investment required to produce meat analogs from soy protein fiber is high relative to most value-added food processes. The process is closer to a pharmaceutical environment than a typical food plant.
2. Yield in the soy spinning process is only about 30%. This triples the cost of the principal raw material, soy flour. In addition, 70% of the raw material is converted into a serious and expensive waste disposal problem.
3. Perhaps the most significant factor was that the consumer was not ready for meat analogs and the price/value was not attractive enough to overcome the "imitation" image of the product. Aside from a small percentage of the population that is attracted to meat analogs for either health or religious reasons, the bulk of the population verified that the product lacked "a reason for being."

It is not unusual for scientists and engineers to find solutions to nonexistent problems. Solutions seeking problems in the food industry are no more prevalent than in other industries. This in fact may not be the case in the long run. One needs only to draw an analogy to margarine to see the possibilities. Margarine struggled for many years as a low-priced alternative to butter. As the technology developed, the product quality was continually improved. The real success of margarine resulted from combining a high quality product with the correct market positioning. Today, margarine is consumed largely because it is perceived to be a healthier product than butter. The price of margarine is very close to butter and, in some instances, higher.

It does not necessarily follow that the same scenario will be true for meat analogs. It is sufficient, at this point, to say that it is technologically feasible to fabricate reasonably good meat analogs from soy protein at a cost that supports a retail price at or below the cost of meat.

XII. TEXTURED VEGETABLE PROTEIN

While the meat analog process was in the midst of commercialization, several companies began to develop an alternative product and technology. Several investigators discovered that thermoplastic extrusion (1,8,11,14) technology could be utilized to form meat-like textured protein products from defatted soy flour. The mechanism is discussed at some length by Harper (10). It is dependent on alignment of protein molecules in the melt stage of extrusion followed by rapid depressurization upon discharge through the extruder die, which results in puffing. This irreversible expansion process permanently sets the protein network with a porous structure that will hold up to three times its weight of liquid. Since the protein has been heat denatured in the extrusion process, which reaches temperatures between 150–200°C, it is insoluble.

Compared to the process for meat analogs, the extrusion process is relatively simple. A typical process flowsheet for textured vegetable protein is shown in Fig. 18.

The extruder is the critical piece of equipment, and a considerable body of technology for the extruder has been developed by various equipment manufacturers (10). Extrusion texturization is characterized by relatively high throughput. An extruder with a 200 HP drive will produce about 5,000 lb/hr (10).

Textured vegetable protein is used primarily as an extender for comminuted meats. Scanning the ingredient declaration of products

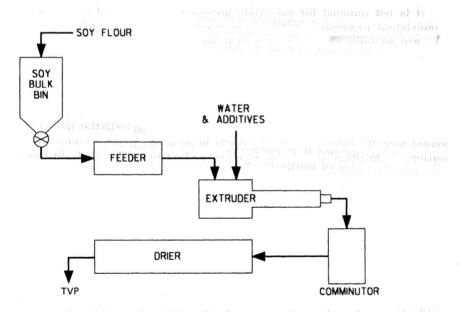

FIGURE 18 Extrusion process for textured protein.

in the supermarket, one would find textured vegetable protein in pizza, chili, spaghetti sauce, t.v. dinners, hamburger patties, and the like. Good quality textured protein can generally be used at levels up to 25% replacement (rehydrated basis) without seriously affecting overall quality. In many instances, especially with less expensive meat blends, quality may actually be improved. On a dry basis, TVP contains approximately 48% protein. When rehydrated at 2:1, it approximates the protein content of ground meat at 16%.

The overwhelming reason for using textured vegetable protein in food products is economic. This is illustrated in the following example.

Textured vegetable protein can be purchased for $0.30/lb or less in dry form.

Uncooked ground beef (50/50 blend) can be purchased for $1.00/lb wholesale.

The cooked yield with added textured protein is better than with 100% ground beef. The textured protein retains some of the meat juices that are normally lost during cooking. The cooked yield from a 70/30 blend of meat/textured protein is about 80% as compared to only 70% for meat alone.

Cost comparison

100% meat	70/30 blend
Basis: 1 lb raw meat	0.7 lb raw meat and 0.3 lb wet textured protein
Cost/lb cooked $1.00/0.7 = $1.43	$[(0.7)(1) + (0.3)(0.1)]/0.8 = $0.91

The savings are quite significant. The application in institutional feeding programs is important and has been a major factor in the textured vegetable protein business since the U.S. Department of Agriculture first permitted its use in the school lunch program in 1971.

REFERENCES

1. Atkinson, W. T., U.S. Patent 3,488,770, assigned to Archer Daniels Midland Co. (1970).
2. Boucher, D. F., J. C. Brier, and J. O. Osburn, *Trans. Am. Institute Chem. Eng.*, 38:967—993 (1942).

3. Bressari, R., et al., *J. Nutrition*, 93:349—360 (1967).
4. Carslaw, H. S., and J. C. Jaeger, *Conduction of Heat in Solids*, Clarendon Press, Oxford, 28388 (1959).
5. Charm, S. E., *The Fundamentals of Food Engineering*, 3rd ed., AVI, Westport, Connecticut, pp. 232—237 (1978).
6. Circle, S. J., and K. Smith, *Soybeans: Chemistry and Technology*, AVI, Westport, Connecticut (1978).
7. Fan, H. P., J. C. Morris, and H. Wakebaun, *Ind. Eng. Chem.*, 40:195—199 (1948).
8. Flier, R. J., U.S. Patent 3,940,495, assigned to Ralston Purina Co. (1976).
9. Goldsmith, et al., Treatment of Soy Whey by Membrane Processes, Proceeding 3rd National Symposium on Food Processing Wastes, pp. 117—149 (1972).
10. Harper, J. M., *Extrusion of Foods, Volume II*, CRC Press, Inc., Boca Raton, Florida, pp. 89—112 (1981).
11. Hayes, L. P., et al., U.S. Patent 3,870,805, assigned to A. E. Staley Co. (1975).
12. Heldman, D. R., and R. Paul Singh, *Food Process Engineering*, AVI, Westport, Connecticut, p. 161 (1981).
13. Lund, D. B., and D. Bixby, Fouling of Heat Exchangers by Biological Fluids, *Int. Cong. Food Sci. Technol.*, 6:27—29 (1974).
14. McAnelly, J. K., U.S. Patent 3,142,571, assigned to Swift & Company (1964).
15. Nagoka, J., S. Takazi, and S. Hotani, *Proc. 9th Intern. Congr. Refrig.*, Paris, 2:4 (1955).
16. Norris, Frank A., *Bailey's Industrial Oil and Fat Products*, D. Swein, ed., 2:175—249 (1964).
17. Page, J. A., and R. C. Dechaine, U.S. Patent 3,498,793 (1971).
18. Page, J. A., and R. C. Dechaine, U.S. Patent 3,558,324 (1971).
19. Perry, R. H., and Chilton, C. H., *Chemical Engineer's Handbook*, McGraw-Hill, New York, New York (1973).
20. Plank, R. Z., *Ges. Kulteind*, 20:109 (in German), cited by A. J. Ede, The Calculation of the Freezing and Thawing of Foodstuffs, *Mod. Refins* 52 (1913/1949).
21. Swein, D., *Bailey's Industrial Oil and Fat Products*, 2:215 (1964).
22. Thulin, W. W., and S. Kuramoto, *Food Technol.*, 21:65—67 (1967).
23. U.S. Dept. of Agriculture, Market Research Report, No. 360, AMS (1959).
24. Westeen, R. W., and S. Kuramato, U.S. Patent 3,118,959 (1964).
25. Wolf, W. J., and J. C. Cowan, *Critical Reviews in Food Technology*, 2:81 ff (1971).

6

Ready-to-Eat Breakfast Cereals

The processing of fluid foods is an aspect of process engineering that has long been the traditional domain of the chemical engineer. The more difficult and less developed unit operations in food processing often occur while handling solids or semisolid materials. It is here that one is more likely to find a need for engineering innovation that demands skills from all the engineering disciplines. The ready-to-eat (RTE) breakfast cereal is a useful case to study in that it is a complex process involving solids.

I. BUSINESS DIMENSIONS

The RTE cereal market, excluding hot cereals, was $4.35 billion in 1986. This amounts to about 1.25% of total food sales in the United States. While the size of the RTE cereal business is impressive, it is interesting to note that it is dominated by a few companies. Margins are high and marketing is very competitive with high advertising costs. In addition, it is one of the more capital intensive sectors of the food processing industry so the cost of entry is high.

The market share by company for 1986 is shown in Figure 1. The top three companies, Kellogg Co., General Mills, and General Foods control 76% of the business.

The industry is very mature but one in which loss of even one share point translates to a sales loss of $43 million or $6.4 million in net income (22).

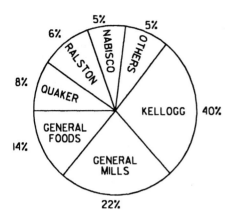

FIGURE 1   Market share for RTE cereal (1986).

II.   PRODUCT CHARACTERISTICS

To understand what appears to be a complex myriad of products with
nonfunctional names, it is helpful to group RTE cereals by generic
type as in Table 1.   RTE cereals fall into five broad categories,
which are familiar to most consumers.   Further insight can be gained
through differentiation by specific product characteristics that cut
across generic categories.

TABLE 1   Market Share by Cereal Type

Type	Share	Typical products
Flaked	35%	Corn Flakes, Wheaties, Raisin Bran
Puffed (gun)	26	Cheerios, Kix, Alpha Bits
Extruded	15	Cap'n Crunch, Chex
Whole grain	8	Rice Krispies
Granolas	7	Nature Valley Granola, Grape-Nuts

Source: From Ref. 1, used with permission.

A. Shape

Shape can be natural as in the whcle grain cereals and granolas or highly fabricated as in extruded and puffed products. Flaked cereals fall between these two extremes.

B. Color

Color can be controlled to some extent through formulation and processing. Natural colors, such as annato or artificial food colors, can be added to the formula. The Maillard reaction can be utilized to develop brown colors by adding reducing sugars and subjecting to heat.

C. Uniformity and Smoothness

Uniformity and smoothness can be controlled for fabricated products that are extruded and puffed. Flaked cereals will appear more natural since irregular blisters occur on the flake surface because of application of radiant heat during toasting.

D. Flavor

Flavor is generally due to a combination of grain type, added ingredients, and heat processing. Grains such as corn, wheat, and rice have distinctly different natural flavors. The use of ingredients such as salt, sucrose, corn syrups, malt syrups, honey, and brown sugar add to the characteristic grain flavor either by flavor enhancement or by forming flavor compounds through the Maillard reaction or complex carmelization reactions. The formation of secondary flavor compounds is complex and must be controlled to produce a consistent product. Variables such as time, temperature, pH, and composition of reactants are key. Ingredients such as corn syrups, malt syrups, honey, and cereal grains are natural products and as such their chemical composition will vary from batch to batch, supplier to supplier, and crop year to crop year. In addition, since the reactants are in a relativly dry slurry, one cannot ignore particle size and shape as variables during the "cooking" process of multiple heterogeneous chemical reactions.

E. Nutrition

Nutrition is becoming more important as consumers develop an awareness of the relationship between health and diet. The addition of vitamins, fiber, and protein to RTE cereals is quite common and tends to generate a specific set of processing and formulation problems. Protein can cause uncontrolled browning reactions resulting in undesirable flavors and/or colors. Fiber can significantly affect texture by acting as an inert diluent in the texturization process. Vitamins are relatively expensive and are applied in minute quantities. The cereal in any randomly selected box must physically contain the vitamin levels stated on the package. Applying small quantities of vitamins, uniformly, over a large quantity of irregularly shaped solids is no easy task. Since many vitamins are heat labile it is not feasible to add them with the other ingredients prior to heat processing. Addition after heat processing is generally the rule.

F. Stability

Stability is a concern for any product that is expected to have a shelf life at room temperature in excess of six months. Human safety is not the issue since RTE cereals are dry and microbially stable products. The concern is in control or prevention of rancidity and the prevention of moisture absorption. Hydrolytic rancidity is controlled by reducing the moisture of the product to 1–3%. Oxidative rancidity is controlled with antioxidants such as BHA and BHT (butylated hydroxy anisole and butylated hydroxy toluene).

G. Texture

Texture in RTE breakfast cereal is critical to product acceptance by the consumer. Crisp and crunchy textures are expected and can be imparted to amorphous, soft cooked grains in a number of ways. Texturizing basically changes the physical shape to alter the surface/volume ratio and thickness, such as in flaking. It can also be achieved by gun puffing or extruding doughs to produce a porous, foamlike structure with greatly reduced apparent density. Other techniques such as shredding are also possible.

III. PROCESSING TECHNOLOGY

To produce a RTE cereal, from either whole grain or cereal flour, several fundamental processing steps must occur. These unit operations are shown in Figure 2.

A. Mixing

Mixing in older plants is batchwise. Continuous mixing systems are available (Readco, Baker-Perkins) and are desirable from the standpoint of a more uniform feed to the process. Uniformity is important

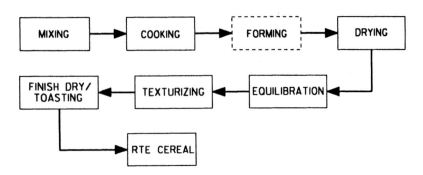

FIGURE 2   Generalized cereal process.

for consistency in flavor development and color, but also for moisture, which is a key process variable in cooking and especially
forming.

B. Cooking

Cooking is essential for flavor development and gelatinization of the
starch to convert it from crystalline form to the amorphous form that
can be readily texturized.

C. Forming

Forming is necessary when dealing with cereal flours as opposed to
whole grains or very coarse grits. Forming extruders of various
designs may be utilized. The key is to minimize shear and prevent
puffing of the extrudate. Uniformity, especially of the critical dimension for heat and mass transfer, is essential. For example,
spheres should be round and not oblate. Rods should have uniform diameter with some tolerance permitted on length.

D. Drying

Drying is critical since moisture content of the formed pellets must
be controlled within a narrow range to ensure uniform rheological
properties for consistent texturization. Nonuniformity can result in
difficulty in controlling texture and bulk density of the product.
The consequences of widely varying bulk density are significant.
Package volume is fixed. If bulk density is too low, it may be impossible to put sufficient product in the package to meet declared
weight. If bulk density is too high, the box will only be partially

filled and will be unacceptable to the consumer. In either instance the product would be rejected and subsequently reprocessed at added cost.

E. Equilibration

Equilibration is often accomplished while conveying or as the product is surged in bins prior to texturizing. The need for equilibration depends on the severity of the moisture gradients generated in the drying operation. Uniformity within and between pellets is important in controlling the texturization process, which does not deal with bulk average properties but rather "sees" individual pellets. This is a major and critical difference between processing fluids which possess "bulk average properties" and discrete solids.

F. Texturizing

Texturizing is physically accomplished by flaking, puffing, or extruding. These operations will be discussed in more detail later.

G. Finish Drying or Toasting

Finish drying or toasting reduces the moisture content to the stable range of 1—3%. It also can be utilized to impart surface texture and toasted flavor by application of high-temperature radiant heat.

The technology can be further differentiated by comparing the generic processes for flaked, puffed, and extruded products in Figure 3.

The choice of raw material will dictate the process for manufacture of a flaked product. The use of cereal flours rather than whole grains offers several advantages such as:

Flexibility in formulation
Uniformity in pellet size and rheological properties
Better control of the cooking operation
Reduced cooking times

The use of whole grain is still prevalent, especially for products with long established brand identity. Once a brand is established it is very difficult to change the process. This is not due as much to the technical issues, but rather more constrained by marketing considerations. The rule is "Thou Shalt Not Change the Product." Even improvement is change and is often difficult to implement. Residence times in batch cookers are 1—2 hours, whereas an extruder can cook flour in a matter of minutes. The extruder operates at temperatures of 130—180°C and moistures of 25—30%, whereas a

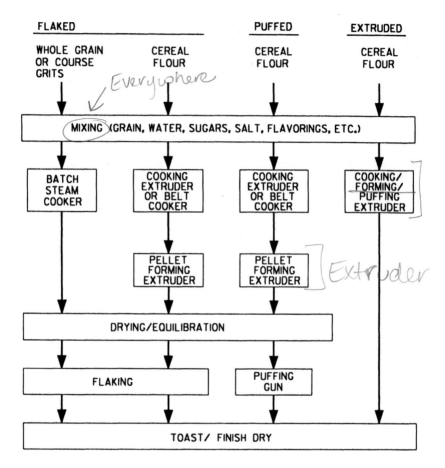

FIGURE 3   Comparison of flaked, puffed, and extruded processes.

batch steam cooker will operate at 110–120°C and moistures of 30–50% (7). Cooking involves more than starch gelatinization. Many chemical reactions, including Maillard, will occur and since the kinetics are not known it is difficult to scale from a batch cooker with heat transfer from condensing steam to a cooking extruder with heat transfer by conduction and shearwork by the extruder screw. In addition, there is a substantial size difference between whole grains such as corn, wheat, and rice and flours. This introduces another uncontrolled variable into the scale-up problem.

It is interesting to compare the puffed and extruded processes. Products can be made by either process that are visually very similar. The extruder appears to offer a single piece of equipment that can combine cooking, forming and puffing into a single step and also eliminate an intermediate drying step. Why then are such a large percentage (26%) of puffed cereals produced by gun puffing rather than the more streamlined process? There are several reasons:

1. Historically, gun-puffed cereals were developed prior to the use of plastic-type extruders in the food industry. Forming extruders such as Macaroni presses or pellet mills were the only machines available to process engineers. These operate at lower temperatures and pressures than are required for puffing.

2. The traditional puffing process of cooking, forming, drying and gun puffing has established a line of products with specific organoleptic properties of taste, texture, and appearance. Of these attributes, texture is the most difficult to reproduce if one desired to make the same products by extrusion puffing.

3. Gun-puffing technology has become quite sophisticated and a substantial capital investment is in place. It is unlikely that this would be abandoned simply to streamline the process. Current gun-puffing systems are highly automated with minimal direct labor and consistent high quality. Displacement of current processes by extrusion puffing would not be economically attractive. It is not likely that extrusion puffing will displace traditional gun puffing for established products. However, new extrusion puffed products continue to be developed and will grow in number.

IV. GUN PUFFING

The gun-puffing process is used to manufacture 26% of the RTE cereal, which translates to approximately $1 billion in sales. Well-known brands such as Cheerios with sales of over $200 million in 1983 (1) command large market shares.

What specifically is gun puffing and how does it work? To understand the physics, consider a single, homogeneous dough pellet which contains approximately 14—18% moisture as depicted in Figure 4.

The dense pellet is isolated physically from its surroundings in the closed box. As heat is applied externally, moisture evaporates from the pellet and raises the pressure within the box. Temperature within the pellet rises as it seeks thermodynamic equilibrium with its environment. Since the environment consists of saturated

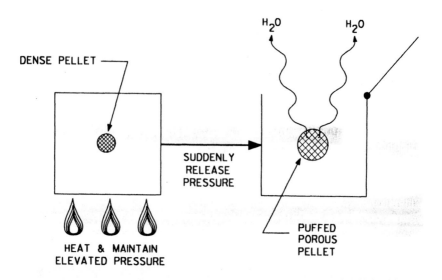

FIGURE 4 Puffing phenomena.

steam there is only one degree of freedom. If the pressure is allowed to rise to 100 psig, the pellet temperature will approximate the saturation temperature of steam at 100 psig (338°F). At this temperature the dough pellet becomes plastic.

When pressure is relieved by opening the box suddenly, the pellet is exposed to a new environment at a much lower pressure. Moisture within the pellet vaporizes rapidly as the pellet seeks a new equilibrium. In so doing, the gas causes the plastic dough to expand. Rapid expansion results in cooling of the pellet and the starch matrix is set irreversibly in a foamlike structure. Typical expansion is three times pellet diameter. For a spherical pellet with volume proportional to (diameter)^3, volume is increased to 27 times the original.

Moisture level in the dough pellet prior to heating is a key variable. If moisture is too high, there is a tendency to explode the pellet rather than maintain a controlled puff. If moisture is too low, the pellet may become charred on the surface since the steam can be easily superheated if moisture is not being vaporized from the pellet.

The first commercial puffing process utilized batch puffing guns. A schematic of a batch puffing gun is shown in Figure 5. Dimensions were approximately 6 inches in diameter and 30 inches in length. Pellets were loaded with a moisture content of approximately

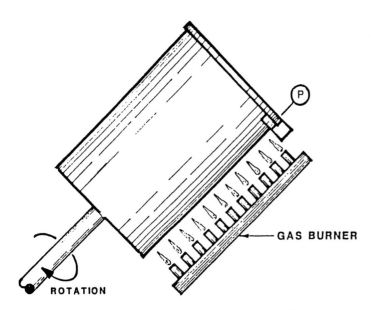

GAS BURNER

ROTATION

FIGURE 5   Batch puffing gun.

12%. Cycle time varied from 5–7 minutes, and the "shot" was made
when pressures were >100 psig (17). Batch puffing exhibited true
transient behavior. There was variation within a batch of pellets
from a single gun due to nonuniform heating. For a single gun
there was variation from shot to shot and there was variation from
gun to gun. The batch guns suffered from high maintenance costs.
Consider the shock frequency for a typical gun. For a gun operated on a 7-minute cycle, 20 hours/day and 6 days/week, this
amounts to over 50,000 thermal and mechanical shock cycles in a
year. These factors provided the incentive to develop a continuous process that would operate at steady state and produce a uniform product in such a manner that the equipment did not self-
destruct. The concept of continuous puffing is depicted in Figure 6.
    The challenge was significant since several pieces of equipment
had to be invented to operate in a production environment with the
constraint that the product from continuous puffing must emulate
the batch-puffed product within the discriminatory range of the
average consumer. In taste panel terms, this means no difference
in a triangle test. —taste panel
    The feed mechanism must be capable of continuously feeding pel-
lets into a vessel with pressures of more than 100 psig without

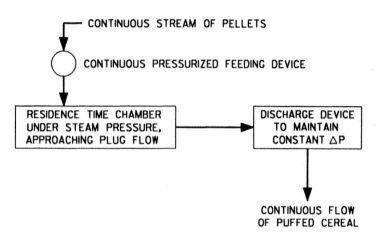

FIGURE 6   Continuous puffing concept.

plugging or sticking. Residence time should be adjustable to accept pellets of varying moisture to develop a minimum pressure for acceptable puffing. Residence time distribution should be as close to plug flow as possible since back mixing would introduce undesirable variability. Each pellet should ideally have the same thermal history. Lastly, the discharge device must permit product that is physically expanding to exit freely without loss of back pressure in the residence time chamber.

Several continuous puffing machines are described in the patent literature. Two of the devices are particularly creative and it is interesting to compare them to see how independent inventors resolved the technical problems.

Haughey (8) of General Foods shows a device (Figure 7) which feeds cereal pellets through a dual-valve chamber to a series of steam-jacketed screw conveyors. High-pressure steam is injected at a point near the fixed discharge nozzle which operates at sonic velocity or choke flow. The high pressure is maintained throughout by providing a purge at the feed end of the chamber. Residence time is controlled by the length of the tube and the rotational speed of the conveying screw.

Tsuchiya (26) of General Mills avoided mechanical-screw conveyance and provided capability for radiant heating as well as superheated steam. The device is shown schematically in Figure 8. Pellets are fed through a rotary feed valve, of special design, utilizing superheated steam as the conveying medium. The feed valve is shown schematically in Figure 9 (23). The residence time chamber

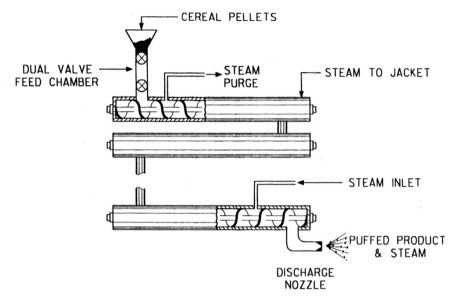

FIGURE 7 General Foods puffing gun (U.S. Patent 2,622,985).

is an elongated rotating pipe which is connected to the feed valve and discharge nozzle through rotary unions. Thus the feed and discharge devices remain stationary as the pipe rotates. Flow through the nozzle is choke and residence time is controlled by length of pipe, angle, and rotational speed. Temperatures in excess

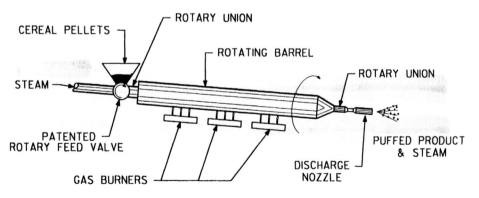

FIGURE 8 General Mills puffing gun (U.S. Patent 3,231,387).

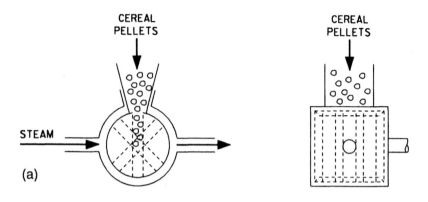

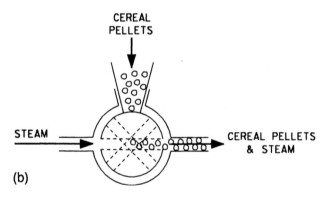

FIGURE 9   Rotary feed valve (U.S. Patent 3,288,053).

of the internal steam temperatures may be applied to the pipe wall by means of external heat sources such as gas burners. This will keep the inner pipe wall dry to prevent sticking and also superheat the pellet surface to enhance desirable browning reactions.

Both devices provide steady state operation under near plug-flow conditions and represent two creative and practical solutions to a difficult processing problem.

Both devices use steam as the heat transfer medium rather than air. The heat transfer coefficient for condensing steam is two orders of magnitude greater than that for air. What effect does this have on the puffing process?

The batch gun cycle time ranges from 5 to 7 minutes. The mode of heat transfer is complex involving convective heat transfer from the air to the pellets, conductive heat transfer from the metal surface, and radiation to the air and the pellets. It is possible to obtain an approximate solution to this problem by applying the unsteady state heat conduction equation. Consider a spherical unpuffed pellet with the physical properties

R = radius of sphere = 3/32 inch = 1/128 ft

k = thermal conductivity of pellet = 0.15 BTU/hr ft °F

ρ = pellet density = 87.2 lb/ft^3

C_p = pellet heat capacity = 0.44 BTU/lb °F

For the batch gun with air as the heat transfer medium assume a heat transfer coefficient h = 0.88 BTU/hr ft^2 °F (9) and that the air surrounding the pellet is at a temperature of 500°F.

The dimensionless numbers that are important in this physical situation are the Biot number N_{Bio} and the Fourier number N_{Fo}:

$$N_{Bio} = \frac{hR}{k} \quad N_{Fo} = \frac{\alpha t}{R^2}$$

where α = thermal diffusivity = $k/\rho C_p$. Graphical solution to the heat conduction equation can be found in many engineering textbooks (3,9). The dimensionless temperature is given as a function of N_{Fo} for various values of N_{Bio} (Figure 10).

The Biot number in this case is N_{Bio} = 0.046, and the dimensionless temperature of interest is when the center of the pellet reaches the saturation temperature of steam at the shot pressure of 100 psig (335°F) for which $(T - T_M)/(T_0 - T_M) = 0.38$; where T_M = ambient air temperature; T_0 = initial temperature. From Figure 10, the Fourier number is estimated to be 3.0. The time required to reach this temperature is calculated for $N_{Fo} = \alpha t/R^2 = 3.0$ and is 2.8 min.

For the continuous puffing gun, the heat transfer medium is actually condensing steam on the pellet surface. The heat transfer coefficient for condensing steam in a sphere is not readily available. However, to compare the two processes, it is adequate to calculate the coefficient from a correlation for filmwise condensation of a vapor on a horizontal tube. Several correlations exist. The one selected here applies for condensation on a tube of outside diameter D (2).

$$h = 0.725 \left(\frac{k_f^3 \rho_f^2 g \, \Delta H_{vap}}{\mu_f D \, (T_d - T_0)} \right)^{1/4}$$

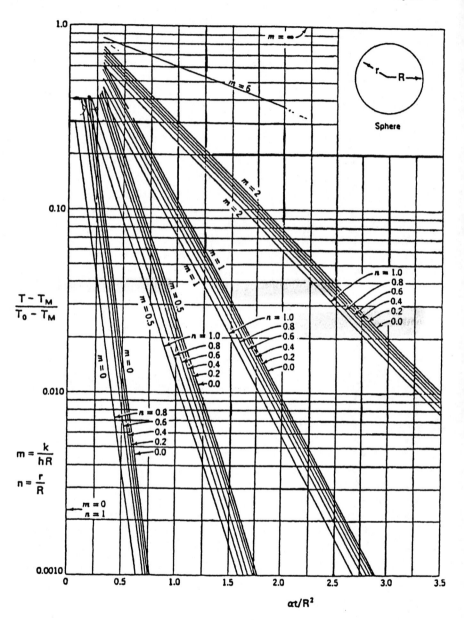

FIGURE 10   Unsteady-state temperature distributions in a sphere. (Adapted from Ref. 9.)

where

k_f = thermal conductivity of condensate = 0.393 BTU/ft °F hr
ρ_f = density of condensate = 56 lb/ft³
ΔH_{vap} = latent heat = 900 BTU/lb
μ_f = viscosity of liquid = 0.45 lb/ft hr
$T_d - T_0$ = temperature drop through condensate film on surface
g = 4.18 × 10⁸ ft/hr²

assuming a temperature drop through the condensate film of 20°F

h = 3444 BTU/hr ft³ °F

Then the Biot number is

$$N_{Bio} = \frac{3444(1/128)}{(0.15)} = 179$$

and assuming a steam temperature of 425°F, the dimensionless temperature is 0.25, and from Figure 10 the Fourier number is 0.18. Then

$$\frac{\alpha t}{R^2} = 0.18$$

from which

t = 10 sec

In this case, the Biot number is so large that the temperature drop through the condensate film is not critical.
What should be concluded from these calculations?

The heat transfer process in an environment using air as the heat transfer medium is rate limited by convective transfer from the air to the pellet. This can be seen from the low Biot number of 0.046.

The large heat transfer coefficient for condensing steam reverses this process so that conduction within the pellet, rather than convective heat transfer is the rate limiting step. This can be seen from the large Biot number of 179.

Injecting steam in a batch puffing gun would shorten the cycle time by increasing the heat transfer coefficient in the Biot number. This has been done in practice.

Residence time required in a puffing gun using condensing steam as a heat transfer medium will be determined by the geometry and characteristic diameter of the pellet.

V. EXTRUSION PUFFING

As indicated in Table 1, extrusion puffed cereals represent a significant share of the RTE market. They also employ some of the most recent technological developments in the industry. Referring to Figure 3, it is clear that combining the cooking, forming, and puffing operation into a single extruder and eliminating the intermediate drying step would have significant capital implications. This is, in fact, one of the primary reasons for the growth of extrusion puffed products. This technology also lends itself to unique products which cannot be readily manufactured by the more traditional techniques of gun puffing or flaking.

The fundamental difference between extrusion puffing and extrusion forming is in the severity of conditions within the extruder barrel. In a puffing operation, extrudate temperature are > 100°C so that moisture in the extrudate will be superheated at atmospheric conditions. As extrudate exits through the extruder die and passes from high pressure to atmospheric pressure it puffs due to rapid vaporization of the moisture in the extrudate.

In a forming operation, temperatures are maintained less than 100°C through proper screw design and cooling of the extruder barrel and/or screw to prevent puffing.

There are several designs of extruders for producing puffed cereals in the patent literature (11,24,29).

VI. GENERIC CORN FLAKE PLANT

A. Description of Plant

It was mentioned previously that margins for RTE cereal were high and processing is capital intensive. Using data and information in the public domain it is possible to develop a first-order capital estimate for a generic corn flake plant as well as a product cost estimate.

The ground rules are as follows:

1. Plant will be sized to produce 1.5 MM (million) cases/year of corn flakes with 24 boxes per case and 12 oz declared weight per box.
2. Plant will operate 16 hours/day, 5 days/week, 50 weeks/year.
3. Operating efficiency is 80%.
4. Product shrink (moisture-free basis) is 5%, and packaging shrink is 2%.
5. Product will be packaged in bag-in-box, and 24 boxes placed in a case.

6. Utilities in the form of electricity, steam, air, and so on, are available on site.
7. A building is available and requires no major modification. This includes processing area and warehouse space.

The process specifications are as follows:

Flow sheet (Figure 11).
Corn flour (or fine grits) with a bulk density of 34 lb/ft³ initially at 12% moisture is blended with sugar, salt, and malt flavoring in a batch system. Water is added to bring the moisture content to 28% and the mixture is cooked and formed into pellets in a cooking extruder at pressures and temperatures low enough to prevent puffing.
The corn flour is pneumatically conveyed from existing bulk bins to the batch mixing system. Preliminary testing by vendors of pneumatic systems indicates a conveying velocity between 4,500 and 5,000 ft/min.

Pilot plant studies have been conducted on a brand "X" cooking extruder with the following geometry and operating conditions:

Diameter, 3.0 in.
Thread depth, 0.375 in.
Thread length, 26.8 in.

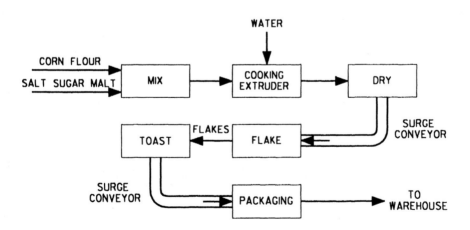

FIGURE 11   Generic corn flake process.

Thread pitch, 1.25 in.
Land width, 0.25 in.
Die pressure, 1500 psig
Screw speed, 32 rpm
Discharge temperature, 190°F
Feed temperature, 90°F
Moisture of extrudate, 28%
Production rate at 28% moisture, 152 lb/hr

To maintain product quality and identity, extrusion temperature
and pressure must be maintained at 190°F and 1500 psig, respectively. Pellets are dried from 28% moisture to 15% moisture and
cooled on a conveying belt. Dried (15%) cooled pellets are flaked
on flaking mills. Flakes are dried and toasted to 3% moisture. Product is cooled, then packed bag-in-box, 12 oz declared weight, 12.25
oz target wt, 24 boxes per case.

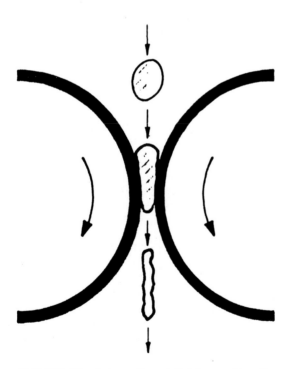

FIGURE 12   Schematic of flaking roll action on cereal pellet.

Some equipment that is adequately sized for this process has already been priced FOB plant. This includes the following:

Pellet dryer, 5 ft × 5 pass, 20 ft long, $560,000
Toasting oven, 2 zone; 4 × 20 ft, $750,000
Packaging equipment (vertical form-fill machines, bag-in-box inser-
ters, and case-packers), $1,600,000
Miscellaneous conveyors and support equipment (including mixers and
belt cookers), $2,000,000
Flaking rolls with a rated capacity of 20–25 lb/min for pellets with
a moisture content of 12–18% require a 60-HP drive and cost
$85,000 each
The pneumatic conveying system (Figure 14) has been conceptually
designed but the size of the conveying lines and blowers must be
determined

Scale-up of the cooking/forming extruder is a critical issue in the
design of the process. The full-scale plant must manufacture product of the same quality and identity as the pilot plant system. This
is an extremely critical point in process development for food products. Many products that have been successful in test markets or
regional markets have floundered during national roll-out attempts
for any of several reasons related to inadequate scale-up. These
include the following:

Inability to scale-up and produce large quantities
Change in product attributes (negative) as a result of inadequate
scale-up
Poor economics resulting from inadequate scale-up

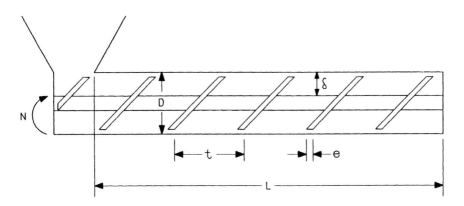

FIGURE 13 Schematic of extruder.

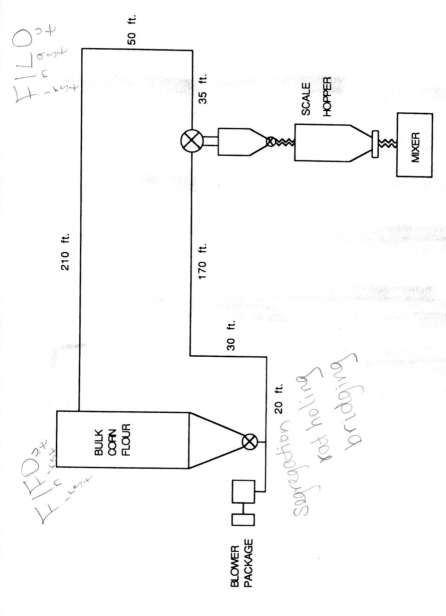

FIGURE 14 Schematic for pneumatic conveying system for corn flour.

For extruders, one must first verify that the full-scale extruder will yield the correct product and then determine actual process rate for costing purposes.

In this example the manufacturer of "brand X" extruders offers a production machine with a nominal capacity of 2,500 lb/hr (at 28% moisture). The geometry and recommended operating speed are

Diameter, 6.5 in.
Thread depth, 1.0 in.
  length, 67.8 in.
  Pitch, 4.25 in.
Land width, 0.5 in.
Recommended speed, 32 rpm
Available coolant temperature, 60°F

This machine is priced at $150,000 FOB plant equipped with appropriate motor specified by customer.

Utility costs are as follows: electric, $0.06/kw hr; natural gas $0.60/therm ($10^5$ BTU). Excluding the flaking rolls and extruders, the smaller electrical motors in the process will total 250 HP (conveyors, feeders, blowers, pumps, etc.).

Typical ingredient costs are as follows (at 12% moisture, except malt syrup):

Corn flour $ 8.00/cwt
Sugar $24.00/cwt
Salt $ 3.00/cwt
Malt syrup $40.00/cwt

Packaging material costs are as follows:

Box $0.50/1,000 in.^{2}
Film $0.15/1,000 in.^{2}
Case (for 24 boxes) $0.48

The hourly labor rate, including fringe benefits is $17.40/hr. In food plants of this type, which are not highly labor intensive, fixed overhead is generally comparable to direct labor burden.

Depreciation is figured at 15 year, straight line.

B. Solution Procedure

Determination of capital and product cost requires the following information:

Material balance
  Ingredient cost
  Packaging material cost
  Labor (direct and indirect)
  Sizing of unspecified equipment
    Extruders
    Flaking mills
    Pneumatic conveying lines and blowers
  Utilities

Material Balance

From the nutritional information and ingredient declaration listed on
corn flake packages available at any grocery store, one can determine a generic formula with sufficient accuracy for a first-order estimate (±30%).

A 1-ounce serving (28.4 g on dry basis) contains the following:
protein, 2 g; carbohydrate, 25 g; sodium, 280 mg (equivalent to
706 mg NaCl). Further, the carbohydrate fraction is broken down
as follows: starch, 23 g; sucrose and other sugars, 2 g.

Thus typical corn flakes would contain

Protein	7.0%
Starch	81.0
Sugar	7.0
Salt	2.5
Other (malt, corn syrup solids, vitamins)	2.5
	100.0%

All the starch and protein are contained in the milled corn. Salt
was estimated from the declared sodium content per 1-ounce serving,
and malt flavoring was assumed to make up the balance. The estimated formula then would be

Milled corn	88.0%
Sugar (sucrose)	7.0
Salt	2.5
Malt flavoring	2.5
	100.0%

Basis is 1 lb product at 3% moisture.

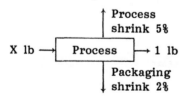

On a moisture-free basis,

$$X = \frac{1.0(0.97)}{(0.98)(0.95)} = 1.04$$

Converting to 12% moisture:

$$\frac{1.04}{0.88} = 1.18 \text{ lb feed/lb product}$$

Since a case contains 24 packages (12.25 ounces/box represents a 1/4-ounce overfill on each package):

$$1 \text{ case} = 24 \frac{(12.25)}{(16)} (1.18) = 21.68 \text{ lb of ingredients}$$

Ingredient costs are available from sources as accessible as the *Wall Street Journal* or from specific ingredient brokers. Typical costs are (for simplicity, it is assumed that prices of ingredients are on the basis of 12% moisture):

Corn flour $ 8.00/cwt
Sugar   24.00/cwt
Salt   3.00/cwt
Malt   40.00/cwt

Thus, the ingredient cost per actual case is $2.12.

Packaging

Packaging material cost for bag-in-box is 0.65 cents per 1,000 in.^{2}. The case cost is $0.48.

For a 12-ounce box, typical package configuration would require 300 in.^{2}. The bag-in-box would then be

$$\text{Cost} = \frac{24(300/1000)(\$0.65)}{0.98} = \$4.78$$

The factor 0.98 takes into account loss of packaging material in the stated 2% packaging shrink. The packaging material cost is then $5.26 per case of 24 (including the case).

Total cost (ingredient + package) = $7.38 case

It is interesting to note the relationship between ingredient and packaging cost!

Other costs such as labor, utilities, fixed overhead, and depreciation will be estimated when the process design is complete.

Process Design

Referring to Figure 11 and the equipment price list, the only operations requiring sizing prior to costing are flaking mills and cooking extruders. Usually drying, toasting, cooling and packaging equipment must be selected, but these have been specified for purposes of this example.

Flaking

The purpose of flaking is to add texture to the cereal by drastically modifying the surface-to-volume ratio and thickness of the pellet. This is accomplished by passing a viscoelastic pellet between a set of closely spaced rolls running at differential speeds (as shown schematically in Figure 12). The result is similar to calendaring in the plastics industry. The cereal pellets must be "plastic" so that they do not shatter or break. Moisture must be uniform to avoid stresses in the flake.

A typical flaking mill will have roll dimensions on the order of a 20-in. diameter × 30-in. length. Since these rolls are only supported on the ends, deflection is an issue if the rolls are too long. This would result in flakes of nonuniform thickness which would be organoleptically objectionable. Thus, there is a practical limit on length and the only effective scale-up method is to add more flaking mills.

Capacity of a flaking mill will ultimately be limited by the ability to feed pellets to the nip of the rolls as fast as possible without producing an excessive number of "doubles." As the name implies, a double occurs when two pellets are flaked in an overlapping fashion to create a fused double flake. This is unacceptable from a product quality standpoint since the resulting flake would be harder and thicker than desired. Several ingenious methods have been devised by both equipment and cereal manufacturers to maximize throughput on flaking mills. Capacity is determined also by the rheological properties of the pellet, which will vary depending on the grain type (corn, wheat, rice, etc.). For corn-based pellets, capacities of 20–25 lb/min at 12–18% moisture on a 20 × 30 in. flaking mill with 60-hp drive are quoted (12). A mill of this capacity would cost about $85,000.

The ground rules specify that the plant will operate 16 hours/day, 5 days/week, 50 weeks/year and must produce 1.5 million cases/year. Then the actual production design rate will be given by

$$\frac{1.5 \times 10^6 \text{ cases/year}}{(16)(5)(50)(0.8)} = 469 \text{ cases/hour}$$

Note the factor of 0.8 to compensate for operating efficiency. Flaking mills are such that it is wise to assume conservative feed rates when the manufacturer provides a range. It is always advisable to have experimental data, since throughput is not predictable from first principles. In this case, it is assumed that the flaking rate will be 20 lb/min at a moisture of 15%. Then,

$$\text{Cases per flaking mill} = \frac{20(0.85)(60)}{(0.97)(19.08)} = 55.11 \text{ cases/hr/mill}$$

The factor 0.97 converts bone dry solids rate to finished product moisture of 3%. The value 19.08 represents the bone dry weight of ingredients needed to yield a case of bone dry product and includes process shrink. Then

$$\text{Number of flaking mills} = \frac{469 \text{ cases/hr}}{55.11 \text{ cases/hr/mill}} = 8.5 = 9 \text{ mills}$$

Flaking mills require maintenance which could take them out of operation. It is not unusual to provide at least one spare mill to prevent loss of production capacity if a mill should go down. For design purposes, 10 flaking mills are then specified.

Extruders

The scale-up criterion for the cooking/forming extruder is that the die pressure must be 1,500 psig at discharge temperature of 190°F to maintain product identity. Several questions must then be answered to complete the scale-up.

1. Will the proposed extruder meet the design criteria? If so, at what rpm?
2. What will be the throughput at this rpm? (This determines the number of extruders.)
3. What is the power requirement? (This determines the motor size.)

Several sources (6,13,20,25) of information are available for estimating the production rate, power consumption, and so on, of extruders.

To use any model for extruder design, a detailed rheological model of the extrudate must be available. The literature (6,30) describes how this information may be obtained. In general, food doughs are found to be pseudoplastic (shear thinning) and have a consistency which is highly dependent on composition and temperature. For this example, the following model for viscosity provides a reasonable description of the rheological behavior of the cereal dough:

$$\tau = m\dot{\gamma}^n$$

where τ is the sheer stress, $\dot{\gamma}$ the sheer rate, m the consistency, and n the flow index. In this case, n = 0.36 by measurement of viscosity with a rheometer. The consistency, m, may be estimated through

$$m = 0.73 \exp(-14 M) \exp(7884/T)$$

where M is the dry basis moisture and T is the absolute temperature °R.

Thermal properties of the dough may be calculated from the composition of the dough (21). For this particular example, heat capacity, C_p, = 0.5 BTU/lb °F; thermal conductivity, k, = 0.15 BTU/ft hr °F.

The simplified models (13,15) proposed in the literature for this type of extruder will be used. The operating characteristics of the full-scale extruder may then be calculated. From the simplified models, the output and power of the extruders may be estimated from the following equations. The output of the extruder is given by,

$$Q_n \simeq \frac{\pi ND(t-e)\delta}{2} - (t-e)\delta\ v_r$$

where Q_n is the net volumetric output, N the rotational rate of the screw, D the screw diameter, t the screw pitch, e the screw land width, δ the thread depth, and v_r the average velocity of reverse flow. These terms are defined further in Figure 13. The average reverse flow velocity may be estimated from the following dimensionless equation:

$$f \simeq \frac{7}{N_{Ref}}$$

where

$$f = \frac{D_{eq}\ P/4\ L_{eq}}{(1/2)\rho v_r{}^2}$$

$$N_{Ref} = \frac{D_{eq}{}^n\ v_r{}^{2-n}\ \rho}{\dfrac{m}{8}\dfrac{(6n+2)^n}{(n)}}$$

where ρ is the dough density (assumed to be 87.2 lb/ft³). D_{eq}, the equivalent screw diameter, and L_{eq}, the equivalent screw length, are defined by

$$D_{eq} = 2\left[\frac{(3n + 1)}{[2\pi(2n + 1)]}\right]^{1/(3+n)} [(t - e)\delta^{2+n}]^{1/(3+n)}$$

$$L_{eq} = \frac{\pi DL}{t}\sqrt{1 + \frac{(t)^2}{(\pi D)}}$$

where L is the screw length.
The power consumed by the screw is given by

$$P = P_s + \frac{P[\pi ND(t - e)\delta]}{2}$$

where P is the total power consumption and P_s is the power consumption due to shear. P_s may be calculated from the dimensionless expression

$$N_{po}\frac{1}{N_{Rep}}$$

where

$$N_{po} = \frac{P_s}{\rho N^3 D^4 L} \quad \text{and} \quad N_{rep} = \frac{\rho\delta^n(ND)^{2-n}}{m\pi^{2+n}}$$

For the initial and final extrudate temperatures of 90°F and 190°F respectively, and die pressure of 1,500 psig, these equations predict that the full-scale extruder will have the following output and power dissipation rate (assuming an average consistency): output = 3,700 lb/hr; power = 124 hp.

If the extruder was not cooled by an external jacket, all of the power dissipated would appear at the extruder discharge as an increase in extrudate temperature. Without cooling, the extrudate temperature is calculated as follows:

$$T_{final} = P/Q_n\rho\, Cp + T_{initial} + 260°F$$

It is clear that significant quantities of energy must be removed through the cooling jacket to maintain an extrudate temperature below 190°. Without cooling, undesirable puffing of the pellets will occur. The heat that must be removed by the jacket may be estimated by

$$Q_j = P - Q_n \rho C_p (T_{final} - T_{initial})$$

$$Q_j = 129,000 \text{ BTU/hr}$$

The rate of heat removal through the barrel walls is given by

$$Q_j = UA_j \Delta T_{LM}$$

where

U $\quad$ = overall heat transfer coefficient
A_j $\quad$ = jacket area (assumed to be 90% of barrel area)
ΔT_{LM} = log mean temperature difference between jacket and extrudate

There is very little published data for estimating the convective heat transfer coefficients for extruders. One source (15) gives the following dimensionless equation for estimating the internal coefficient:

$$Nu = 2.2 \text{ Br}^{0.79}$$

where

$$Nu = \frac{h\delta}{k} \quad \text{(Nusselt number)}$$

$$Br = \frac{m(\pi ND)^{n+1}}{k\Delta T_{LM}\delta^{n-1}} \quad \text{(Brinkman number)}$$

where h is the convective heat transfer coefficient of the extrudate and k is the conductivity of the extrudate.

Pilot plant experimentation such as that described in the chapter on scale-up would provide a basis for estimating this coefficient.

Assuming that the jacket side coefficient has a value of 500 BTU/hr ft^2 °F and the barrel wall is 3/4-in. stainless steel, the overall heat transfer coefficient may be estimated by

$$\frac{1}{U} = \frac{1}{h} + \frac{\Delta x}{k_w} + \frac{1}{500}$$

where Δx is the wall thickness and k_w is the wall conductivity; from which U = 90.7 BTU/hr ft^2 °F.

If 60° water is available, the maximum heat that can be removed by the jacket is given by $Q_j = UA_j\Delta T_{LM} = 86,000$ BTU/hr.

This is less heat removal than required to meet the scale-up criteria of 129,000 BTU/hr. To maintain the final extrudate temperature below 190°F, the rate of energy dissipation per pound of product must be reduced. The alternatives are as follows:

1. Changing extruder geometry, such as the length-to-diameter ratio and the thread depth-to-diameter ratio
2. Reducing extrusion pressure, by increasing the open area of the die
3. Reducing the screw speed

The first choice requires the design and/or selection of another extruder, which may not be available, or redesign of the screw of the available extruder, which is an expensive alternative and requires experimentation. Reducing the pressure reduces energy dissipation per pound, but results in violation of one of the scale-up criteria. The only viable alternative is to reduce the speed of the extruder. This requires a trial and error solution with screw speed as a variable. It is found that at 24 rpm, heat removal capacity balances the heat generation and that output equals 2780 lb/hr, and power equals 83.9 hp.

Since the material balance requires a total of (469)(19.08)/0.72 or 12,428 lb/hr at 28% moisture, the plant will require 4.47, or 5, extruders operating at 24 rpm.

This does not complete the scale-up calculations. In order to meet the discharge pressure requirement of 1500 psig, the number of die openings at the discharge of the full scale extruder must be calculated. The number of die openings required is given by

$$N_{Full} = \left(\frac{Q_{n,Full}}{Q_{n,Pilot}} \right) N_{Pilot}$$

It has been assumed that by meeting the temperature and pressure specifications of the design criteria the process would be properly scaled-up. Other scale-up criteria (6,10,19) have been suggested. They are as follows:

1. Specific energy input
2. Screw shear rate
3. Average total shear strain
4. Screw residence time

These four parameters are given in Table 2 for both sizes of extruder. None of these parameters is maintained as a constant when scaling from the pilot plant unit to the full-scale plant. To accomplish this,

(handwritten margin note: Flaking Design → Flowrate → Power input → Heat Xfer → Die design Scaleup)

TABLE 2   Alternative Scale-Up Parameters at 24 rpm

	Pilot	Plant
Specific energy (kw-hr/lb)	0.0548	0.0226
Shear rate (sec⁻)	13.4	8.17
Total strain	1,210	666
Residence time (sec)	90	81.5

considerable redesign of the full-scale screw geometry, speed, and/
or discharge pressure would be required. If screw speed is the
only design degree of freedom, the screw speeds in Table 3 would
be needed to meet these criteria.

For the first two criteria, the screw speed must be higher than
the maximum speed allowed to meet the temperature control criterion.
This implies that major extruder redesign is required or that per-
fect scale-up may not be possible to accomplish. For this later case,
relaxation of the quality criteria is required. This is not a desirable
course. The product could be reformulated in an attempt to produce
acceptable product at feasible operating conditions of the extruder.
This may not be possible and would certainly require a considerable
product/process development effort.

Pneumatic Conveying

For large-scale food processing operations, pneumatic conveying of
granular solids is used to transport material from bulk storage to
processing areas. Figure 14 shows a typical system, specific to

TABLE 3   Screw Speed Required to Meet Indicated Criterion

Criterion	Pilot plant determination	Speed required (rpm)
Specific energy input	0.0548	305
Screw shear rate	13.4	36
Total shear strain	1,210	None found
Residence time	90	22

the example. In the food processing industry the design of pneumatic systems has evolved through the combined efforts of vendors and users with the accumulation of much data. The effective design of a pneumatic system requires considerable experimental data (usually from the vendor) and practical experience with pneumatic systems. A good working relationship between vendor and user is important in arriving at an optimum design.

Several factors must be specified before a pneumatic system is designed. These include the following:

Quantity of solid to be moved
Physical system configuration, including distances and elbows
Pick-up and conveying velocity, which is based on experience and
    testing by the vendor

The objective of the design is then to determine:

Diameter of conveying line
Blower horsepower

In this example it has been determined that the conveying velocity will be between 4,500 and 5,000 ft/min.

General Procedure

Energy requirements for the system are due to pressure drop in moving both air and solids.

$$\Delta P_{total} = \Delta P_{air} + \Delta P_{solids}$$

Each of these can be broken down further, for example, $\Delta P_{air} = \Delta P_{friction} + \Delta P_{other}$ and ΔP_{other} is pressure loss due to inlet filter, silencer, and line diverters. This information is available from vendors. For this problem, $\Delta P_{other} = 0.5$ psi and $\Delta P_{friction} = K \cdot$ equivalent distance.

Equivalent distance is the sum of the horizontal and vertical distance plus equivalent elbow distance. The equivalent elbow distance is provided by vendors and in this example is assumed to be 15 ft per elbow, which is on the conservative side. K is the pressure drop per 100 ft and can be obtained from a duct friction chart for smooth wall tubing and a roughness correction chart similar to those in Figure 15. Typical K values for the velocities and tube diameters considered here are given in Table 4.

$$\Delta P_{solids} = \frac{E_{total}}{ACFM}$$

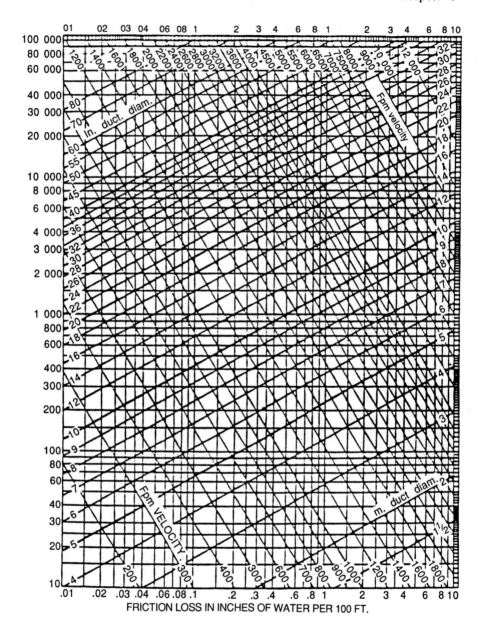

FIGURE 15a   Friction chart for round ducts.   (Adapted from *Fan Engineering*, Buffalo Forge Company, p. 106, 1970.)

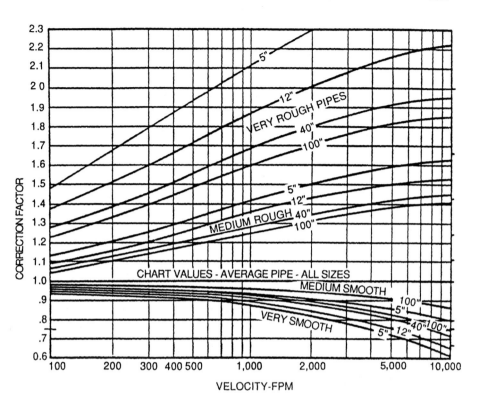

VELOCITY-FPM

FIGURE 15b Roughness correction factor. (Adapted from *Fan Engineering*, Buffalo Forge Company, p. 102, 1970.)

with

$$E_{total} = E_{accelerate} + E_{lift} + E_{friction\ loss\ horizontal} + E_{loss\ in\ elbows}$$

and

$$E_{accelerate} = \frac{\dot{W}V^2}{2g}$$

$$E_{lift} = \frac{\dot{W}h}{60}$$

$$E_{horizontal} = \frac{\dot{W}Lf_s}{60}$$

TABLE 4   K-Values (Pressure Drop Per 100 Ft)

Tubing O.D. (in.)	Tubing I.D. (in.)	Area (sq. ft.)	Velocity (ft/min)		
			4,000	4,500	5,000
3.00	2.83	0.044	0.29	0.35	0.43
3.50	3.33	0.060	0.26	0.30	0.38
4.00	3.81	0.079	0.20	0.26	0.32
4.50	4.31	0.101	0.18	0.22	0.29
5.00	4.75	0.123	0.17	0.20	0.25
6.63	6.36	0.221	0.11	0.16	0.20
8.63	8.26	0.372	0.09	0.11	0.15
10.75	10.35	0.584	0.08	0.09	0.11
12.75	12.00	0.785	0.02	0.06	0.08

$$E_{elbow} = \text{centrifugal force} \times \frac{\text{circumference}}{4} \times \text{friction factor} \times \text{number of elbows}$$

$$= \left(\frac{\dot{W}V^2}{gr}\right)\left(\frac{2\pi V}{4}\right) f_s N$$

$$= \frac{\dot{W}}{2g} V^2 \pi f N = (E_{accelerate})\pi f N$$

where

$\dot{W}$ = mass rate solids, lb/min
V = conveying velocity, ft/min
h = vertical travel, ft
L = horizontal travel, ft
f_s = friction factor for solid = 0.4*
N = number of elbows
g = gravitational constant

---

*For corn grits on steel tubing, f_s = 0.4. [From Jorgenson, R. (ed.), *Fan Engineering*, Buffalo Forge Company, Buffalo, New York, p. 490 (1970).]

and

ACFM = actual volumetric flow rate of conveying fluid

$$= \frac{\pi D^2}{4} V$$

where D is the inside pipe diameter.
To be conservative, a 10% safety factor can be added to ACFM so that

$$\Delta P_{solids} = E_{total}/(1.1)ACFM$$

Calculation Algorithm. A trial and error calculation procedure is used to find a combination of blower and tube size that satisfies the design criteria. Specifically,

1. Pick V and D.
2. Calculate ΔP_{air} from Table 4 and equivalent conveying distance.
3. Calculate E_{total}.
4. From D and V, calculate ACFM and ΔP_{total}.
5. Using operating curves for specific vendor blowers, find horsepower (Figures 16 and 17).
6. Select D and blower to meet design criteria.

For this example, the design criteria are as follows:

Corn flour with density of 34 lb/ft³ and $f_s = 0.4$
Air pick-up velocity of 4,500—5,000 ft/min
Solid/air mass ratio < 20:1 (a good rule of thumb for pressure systems)
System operation with pressure of 7—10 psig (typical blowers have a 12—15 psig operating limit)

Solution.

Calculation of mass rate:

$$\dot{W} = 1.5 \times 10^6 \ cases/year \times \frac{1}{(50 \times 5 \times 16) \ hr/yr} \times$$

$$(21.68)(.88) \ lb \ corn \ flour/case \times \frac{1}{0.8} \ system \ efficiency$$

$$\dot{W} = 8943 \ lb/hr \ corn \ flour$$

Sizing mix/scale delivery system: Assume four batches per hour with 5 minutes allowed to change scale hopper; then, the instantaneous delivery rate $\dot{W}$ is

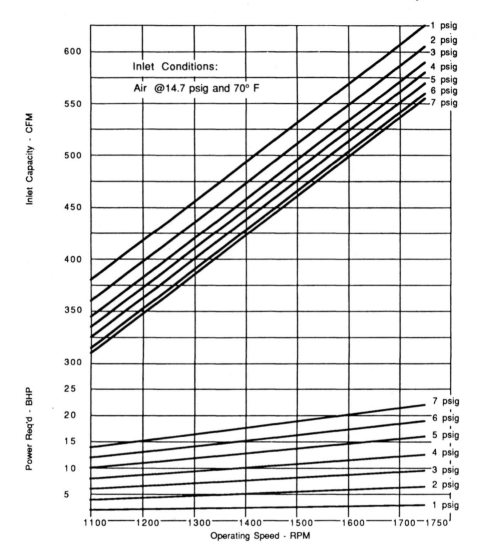

FIGURE 16 Sutorbilt 6MB California Series B Blower. (Adapted
from Sutorbilt Performance Curves, Fuller Company, Compton, CA.)

$$\dot{W} = 8943 \text{ lb/hr} \times \frac{1}{4 \text{ batches/hr}} \times \frac{1 \text{ batch}}{5 \text{ min}}$$

$$\dot{W} = 447 \text{ lb/min}$$

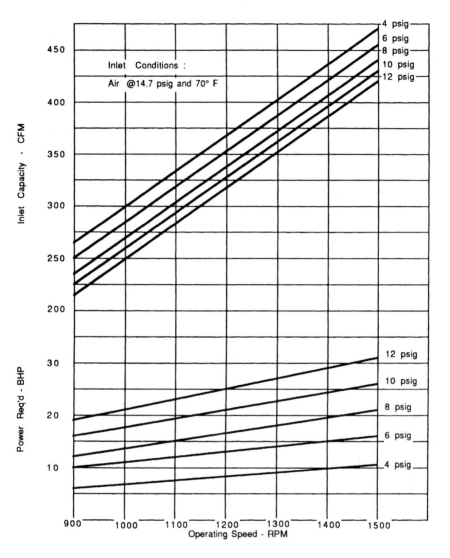

FIGURE 17 Sutorbilt 7HB California Series B Blower. (Adapted from Sutorbilt Performance Curves, Fuller Company, Compton, CA.)

Conveying equivalent distance: From Figure 14, horizontal = 435 ft; vertical = 80 ft; four 90° elbows × 15 ft = 60 ft; total equivalent distance = 575 ft.

With this information, Table 5 is generated by following steps 1 through 4 for velocities of 4,500 and 5,000, and various tube sizes.

Referring to the design criteria, all tube diameters (O.D.) except 4.0, 4.5, and 5.0 in. are eliminated on the basis of allowable pressure drop range (7–10 psig). All of these will satisfy the solids/air ratio <20:1 criterion.

TABLE 5 Pressure Drop Calculation

Tubing O.D.	Material air ratio	Psi drop per 100 ft	(1) Total air loss (psi)	ACFM	(2) Energy to accelerate (ft lb/ min)	(3) Energy to lift (ft lb/ min)
Conveying velocity, 4,500 ft/min						
Friction factor (f_s), 0.4						
3.00	30.2	0.35	2.51	197	39,075	35,760
3.50	21.8	0.30	2.23	272	39,075	35,760
4.00	16.6	0.26	2.00	356	39,075	35,760
4.50	13.0	0.22	1.77	456	39,075	35,760
5.00	10.7	0.20	1.65	554	39,075	35,760
6.63	6.0	0.16	1.42	993	39,075	35,760
8.63	3.5	0.11	1.13	1675	39,075	35,760
10.75	2.3	0.09	1.02	2629	39,075	35,760
12.75	1.7	0.06	0.85	3534	39,075	35,760
Conveying velocity, 5,000 ft/min						
Friction factor (f_s), 0.4						
3.00	27.1	0.43	2.97	218	48,240	35,760
3.50	19.6	0.38	2.69	302	48,240	35,760
4.00	15.0	0.32	2.34	396	48,240	35,760
4.50	11.7	0.29	2.17	507	48,240	35,760
5.00	9.6	0.25	1.94	615	48,240	35,760
6.63	5.4	0.20	1.65	1103	48,240	35,760
8.63	3.2	0.15	1.36	1861	48,240	35,760
10.75	2.0	0.11	1.13	2921	48,240	35,760
12.75	1.5	0.08	0.96	3927	48,240	35,760

[a]Assumes 10% component loss.

With the choices narrowed down to three tube sizes and two velocities, Table 6 is generated from the operating curves for the specific blowers being considered. In this case, they are for different sizes of Sutorbilt blowers (Figures 16 and 17). These can be obtained from manufacturers. Note that 1 psi has been added to calculated pressure drop as a safety factor.

(4) Energy to move horiz (ft lb/ min)	(5) Elbow losses (ft lb/ min)	(6) Total energy for prod (ft lb/min) 2 + 3 + 4 + 5	(7) Psi losses material	(8) Total psi sys drop 1 + 6	(9)[a] ACFM air required
77,778	196,411	349,023	12.33	14.84	216
77,778	196,411	349,023	8.91	11.13	299
77,778	196,411	349,023	6.80	8.80	392
77,778	196,411	349,023	5.32	7.08	502
77,778	196,411	349,023	4.38	6.03	609
77,778	196,411	349,023	2.44	3.86	1,092
77,778	196,411	349,023	1.45	2.58	1,842
77,778	196,411	349,023	0.92	1.94	2,892
77,778	196,411	349,023	0.69	1.53	3,888
77,778	242,482	406,260	12.85	15.83	240
77,778	242,482	406,260	9.28	11.97	333
77,778	242,482	404,260	7.09	9.43	435
77,778	242,482	404,260	5.54	7.71	557
77,778	242,482	404,260	4.56	6.50	667
77,778	242,482	404,260	2.54	4.19	1,213
77,778	242,482	404,260	1.51	2.87	2,047
77,778	242,482	404,260	0.96	2.09	3,213
77,778	242,482	404,260	0.71	1.67	4,320

TABLE 6   Blower Candidates Versus Tube Size (Sutorbilt Series B Blowers)

| Tubing O.D. (in.) | Velocity (ft/min) | | | | | |
| | 3,400 | | | 5,000 | | |
	psi	hp	Type	psi	hp	Type
4.0	10	24	7 HB	11	28	7 HB
4.5	8	22	6 MB		X	
5.0	7	X	X		X	

The 4-in. tubing is selected since it satisfies the design criteria and is lower in cost and less expensive to install than 4.5-in. tubing.   The 7-HB blower equipped with a 30-HP motor will be more than adequate for the specified service.

Drying, Toasting, Cooling

Equipment for the drying, toasting, and cooling operations is often sized by the supplier. Purchase specifications are generally written with a performance guarantee so that financial responsibility for reaching the specified moisture is borne by the supplier once a price is quoted. Generally some form of pilot plant data is required for design verification.   Trials performed at the suppliers location can be misleading since pellets and flakes would be thoroughly equilibrated, thereby eliminating one of the major process variables encountered in a plant environment.   For this example the pellet dryer, toaster, and final cooling have previously been specified.

Packaging

In many food processes, packaging machinery is often the single most expentive item on the equipment list.   This case is no exception.   In this instance, the packaging equipment would consist of two bag-in-box lines rated at 70 cartons/min (each), with associated case packaging equipment to handle 6 cases/min.

C.   Capital Estimate

On the basis of previous calculations and specified equipment the capital estimate can be completed (Table 7).
Both contingency and engineering are significant items in the capital estimate, amounting in aggregate to over 20% of the total

TABLE 7   Capital Estimate

	Cost
Item	
Extruders (5 @ $150,000 ea.)	$ 750,000
Flaking mills (10 @ $85,000 ea.)	850,000
Pellet dryer	560,000
Toasting oven	750,000
Miscellaneous conveyors, support equipment	2,000,000
Packaging	1,600,000
Subtotal	$6,510,000
Installation	
Process	$2,455,000
Packaging	320,000
Subtotal	9,285,000
Engineering, 10%	930,000
Contingency, 15%	1,532,000
Total	$11,747,000

cost. At the early stages of project evaluation, these costs, and in particular, contingency costs, are usually challenged by the financial analysts. As much as one would wish for a capital estimate that is all-inclusive, that will never be the case. Many factors beyond the control of the estimator enter into the design and installation phases so that some contingency is always necessary. The size of the contingency is inversely related to the quality of the capital estimate, which in turn is a function of the amount of engineering design that precedes the actual estimate. One would prefer to have the contingency as low as possible but with minimum risk of being wrong. Taking the time to do enough engineering before final project approval is the only way to achieve this. Unfortunately, time is often more scarce than money and for a variety of reasons capital estimates are often presented with only a minimum of engineering design. The engineer being conservative by nature and in a visible risk position argues for a contingency level in the comfort zone. Finance and marketing in their zeal to present the project in the best possible light and maximize ROI will attempt to reduce the contingency. If one has to fight, it is usually advantageous to pick the time and place. It is better to argue the contingency issue in the early stages of a project than at the 90% completion point when it becomes very critical.

Engineering costs are easier to sustain since they can be developed by detailing specific tasks. In effect, a zero-based engineering estimate can be utilized instead of a straight percentage of the installed capital. There are times when a percentage of installed equipment cost can result in greatly understating actual engineering costs. If a project involves used equipment or retrofitting of an existing facility, there is risk in applying straight percentages. If a significant amount of the equipment is one-of-kind, engineering costs can be disproportionately high. For a custom-designed piece of equipment engineering costs of 20—50% are not unusual. If process design criteria are not comprehensive, there is risk of a protracted start-up with substantially higher engineering costs to do process development work during start-up.

With the capital estimate completed it is possible to finish the product cost estimate.

D. Labor

RTE cereal lines such as this are not labor intensive. While it is not possible to determine exact labor, it should be possible to come within ±30% with some reasonable assumptions. A simple approach is to break the process into work units and assign the labor to each unit.

Thus, the number of people needed for each task is as follows:

Cook/extrude	2
Dry	1
Flake	2
Toast	1
Packaging	4
Warehouse	1
Inspectors	2
Maintenance	1
Relief	1
Supervision	1
	16

Assuming an average hourly labor rate of $17.40/hr, including fringe benefits of 35% (27), the labor cost is $278.00/hr. This is to be assessed against a production rate of 1.5 million cases/year, which amounts to direct labor of $0.74/case. Even if the assumed manning is off by 30% (±5 people), the effect would only be ±$0.23/case.

E. Utilities

Utilities can be lumped into drying, toasting, and electric power for incidental motors, flaking, and extruder cooking and forming.

Drying prior to flaking can be estimated by accounting for latent heat of vaporization. Assuming that the latent heat of vaporization λ = 1000 BTU/lb, that the product reaches 170°F, and that there is a moisture reduction from 28% to 15%, the enthalpy load for drying can be calculated. Sensible heat will be negligible assuming minimal heat loss in conveying between the extruder and pellet dryer. Basis: 1 lb of pellets @ 28% moisture. Amount of water after drying to 15% = X. $X/(X + 0.72) = 0.15$, and X = 0.127 lb. Then

Moisture removed = 0.280 - 0.127 = 0.153 lb/lb wet pellets

and

$$Q_{drying} = (1,000 \text{ BTU/lb water})(0.153 \text{ lb water/lb wet pellets})$$

$$(20 \times 9 \times 60 \times 0.8 \text{ lb/hr wet pellets})(16 \times 5 \times 50 \text{ hr/yr})$$

$$Q_{drying} = 5.287 \times 10^9 \text{ BTU/yr}$$

Assuming 75% dryer efficiency

$$Q_{drying} = 7.05 \times 10^9 \text{ BTU/yr}$$

Toasting involves both sensible heat and latent heat. Product is dried from 15% to 3% and, since toasting occurs, reaches temperatures in excess of 212°F. An assumption of 250°F would be reasonable. Also, it is reasonable to simplify the calculation by assuming the water removal is 0.15 lb/lb product.

Then with a heat capacity of 0.5 BTU/(lb °F)

$$Q_{Toast} = (1000)(0.15) + 0.5(250 - 70)$$

$$= 240 \text{ BTU/lb product}$$

Again assuming 75% efficiency, Q_{Toast} = (240)/(0.75) BTU/lb product × (1.5 × 10^6 cases/yr) (24 × 12.25/16 lb/case) = 8.82 × 10^9 BTU/year. Total Q = (7.05 + 8.82) × 10^9 BTU/year.

With natural gas at \$0.60/therm, this contributes \$95,200/year or \$0.0635/case.

Electric energy costs are as follows:

Incidental motors = 250-HP, so (250)(0.746) = 186.5 kw
Flaking mills at 60 HP each (assume full load and note that only nine flaking mills are utilized, the tenth mill being a spare), so 9 × 60 (0.746) = 402.8 kw
Extruders at 84 HP, so (5)(84)(0.746) = 313.3 kw
Total = (902.6 kw)(\$0.06 kw hr) = \$54.16/hr for an instantaneous rate of 469 cases/hr, this adds \$0.12/case.

F. Total Cost

The total cost per case is then as follows:

Ingredients	$2.12
Packaging	5.26
Direct labor	0.74
Fixed overhead	0.74
Utilities	0.19
Depreciation	0.52
	$9.57/case

Corn flakes sold for about $1.13 at the supermarket in an informal
survey of Minneapolis supermarkets in August, 1985. For a case
of 24, the retail case value was $27.12. A supermarket might have
a margin of 30%, which translates to a cost of $18.98 to the super-
market. The manufacturer's margin of $9.41 must cover transpor-
tation, distribution, sales promotion, advertising and administrative
costs and, of course, some profit. RTE cereals are heavily adver-
tised. Because promotional campaigns on a national level are very
expensive, large gross margins are required to support a value-
added product of this type where there is very little true differentiation between brands.

G. Sugar Coating

RTE cereals are certainly value-added products. Recognizing that
many individuals add sugar as well as milk to their breakfast cereal, the cereal manufacturers developed a new product category
of presweetened RTE cereal.
    Three types of sugar coating are discussed in the patent litera-
ture.
    Frosted coatings are obtained when coating is accomplished by
coating with sucrose solution followed by drying. Sucrose crys-
tallizes during the drying, which results in the familiar frosted appearance.
    Transparent coatings can be obtained by applying molten sugar
with a very low moisture content (3–6%). This process is an extension of candy coating technology (16). It is a difficult process
to control and the molten sugar has a short pot life on the order
of several minutes before carmelization begins. An alternative
process utilizes 1–8% reducing sugars such as invert sugar, glucose, or fructose in a sucrose solution. Drying leaves a clear
glaze that does not become sticky in high humidity (28).
    Crystalline sucrose granules can be physically bound to cereal
by utilizing binders of gelatin, corn starch, or carboxymethylcellu-
lose (18). This has the advantage that the sucrose is never

dissolved and does not alter the physical appearance of the substrate.

To complete this example, consider the manufacture of a frosted, sugar-coated corn flake. This is simply an addition to the corn flake plant that involves coating with sugar solution followed by drying to remove excess moisture. How is the product cost altered by sugar coating? From nutritional labeling information, the starch and sugar content in a 1-oz serving (28.38 g) is 13 g each. Thus, the sugar-coated flake contains 13 g/oz or 46% starch. The flake itself contained 77.5% starch prior to coating. Then the flake represents 46/77.5 = 59% of the coated product. The composition would be 59%, flake; 41%, added sucrose as coating. The composition of the coated product is approximately as follows:

Milled corn	50.1%
Sucrose	45.0
Malt flavoring	3.1
Salt	1.8
	100.0%

The ingredient costs as given previously are as follows:

Milled corn	$ 8.00/cwt
Sucrose	24.00/cwt
Malt flavoring	40.00/cwt
Salt	3.00/cwt

The ingredient costs contribution increases from $2.12/case to $3.49/case. It is easy to see that the price of sugar becomes an important factor in the manufacture of sugar-coated cereals.

VII. SUMMARY

RTE cereals represent a large and important segment of value-added processed foods. Although the first products were developed more than 60 years ago, the technology has increased in complexity over the years and continues to evolve as evidenced by the many new products being produced on extrusion-cooking systems. RTE cereal processing encompasses some very complex heat transfer problems and is a challenging area for additional fundamental research.

RTE cereals are an amazing success story in that they are very contemporary in today's market even though they are old concepts. Today's emphasis on the relationship between health and food speaks to one of the major strengths of RTE cereals. The benefits of these products cannot be judged merely on the value of the ingredients in the box. 'They are a classic example of the concept of value-added products. They deliver quality, convenience, good taste, nutritional value, and in some cases health benefits such as high fiber content at a very low cost per serving.

ACKNOWLEDGMENT

The section on pneumatic conveying was contributed by Bruce Henry (Pillsbury Co.). The extruder scale-up example was contributed by Leon Levine (Leon Levine and Associates).

REFERENCES

1. *Advertising Age*, May 27:32 (1984).
2. Bird, R. B., W. E. Stewart, and E. N. Lightfoot, *Transport Phenomena*, John Wiley and Sons, New York (1960).
3. Charm, S. E., *The Fundamentals of Food Engineering*, AVI, Westport, Connecticut (1981).
4. Choi, Y., and M. R. Okos, Thermal properties of liquid foods—review, ASAE Paper No. 83-6516, presented at Winter Meeting of ASAE, December (1983).
5. Graham, W. R., and B. Grogg, U.S. Patent 3,054,677 (1962).
6. Harper, J. M., *Extrusion of Foods*, Vol. I, CRC Press, Boca Raton, Florida (1981).
7. Harper, J. M., *Extrusion of Foods*, Vol. II, CRC Press, Boca Raton, Florida (1981).
8. Haughey, C. F., and R. T. Erickson, U.S. Patent 2,622,985 (1952).
9. Heldman, D. R., and R. Paul Singh, *Food Process Engineering*, AVI, Westport, Connecticut (1981).
10. Holay, S. H., and J. Harper, Plant protein texturization under varying shear environments, presented at AICHE National Meeting, August (1981).
11. Kelley, E. F., and F. J. Thomas, U.S. Patent 3,458,322 (1969).
12. Lauhoff, personal correspondence.
13. Levine, L., Estimating output and power of food extruders, *J. Food Proc. Eng.*, 6:(1982).
14. Levine, L., and J. Rockwood, Simplified models for estimating isothermal operating characteristics of food extruders, *Biotechnology Progress*, 1:3 (1985).
15. Levine, L., and J. Rockwood, A correlation for heat transfer coefficient in food extruders, *Biotechnology Progress*, 2:2 (1986).

16. Massmann, W. F., E. W. Michael, and W. L. Vollink, U.S. Patent 2,689,796 (1954).
17. Matz, S. A., *The Chemistry and Technology of Cereals as Food and Feed*, AVI, Westport, Connecticut (1959).
18. McKown, W. L., and P. K. Zietlow, U.S. Patent 3,615,676 (1971).
19. Meuser, V. F., B. Van Lengerick, and H. R. Berlin, *Kochextrusion vo Starken*, Starke, 36:194–199 (1984).
20. Middelman, W., *Fundamentals of Polymer Processing*, McGraw-Hill, New York (1977).
21. Mohsenin, N. N., *Thermal Properties of Food and Agricultural Materials*, Gordon and Breach, New York (1980).
22. *New York Times*, April 17:29, 36 (1985).
23. Pertula, H. V., U.S. Patent 3,288,053 (1966).
24. Reinhard, R. D., and R. W. Stephenson, U.S. Patent 3,458,321 (1969).
25. Shenkel, G., *Plastic Extrusion Technology and Theory*, American Elsevier, New York (1966).
26. Tsuchiya, T., G. Long, and K. Haeha, U.S. Patent 3,231,387 (1966).
27. *U.S. Industrial Outlook*, Average wage in cereal breakfast foods industry for 1982, pp. 41–42 (1985).
28. Vollink, W. L., U.S. Patent 2,868,647 (1959).
29. Wenger, J., U.S. Patent 3,117,006 (1964).
30. Wilkinson, W. L., *Non-Newtonian Fluids*, Pergamon Press, New York (1960).

7
Dairy Products

The dairy industry (fluid milk, cheese, ice cream, butter, yogurt, and cream) is a significant part of the food processing industry and of agriculture. Dairy products contribute substantially to the typical diet and represented $45.4 billion in annual sales in 1987 (7). This was about 13% of total food sales that year.

Because of its high nutritive value and relative lack of natural preservatives, milk is highly susceptible to spoilage and can support health-threatening microorganisms, such as salmonella, tuberculosis, and listeria. Thus, sanitation in handling and processing is especially critical. Many of the principles of sanitation and sanitary design in food processing have been developed by the dairy industry.

Likewise, some of the major developments in automation and process control have been first developed for dairy applications. In particular, the concept of cleaning in place (CIP) made possible the construction of large dairies and has now been applied to many other food processes.

Like many other foods, milk generates by-products when it is processed for certain purposes. Fluid whole milk, of course, has nearly the same composition as raw milk, but it is only one of the many products that are routinely made from milk; as the composition is modified, excess amounts of one or more components are generated. These typically differ in value and so can create a challenge in identifying effective uses or harmless means of disposal.

Finally, milk and milk products represent significant sources of important nutrients, especially protein and calcium. While most Americans consume more protein than necessary, calcium is significantly lacking from the diets of many people, especially women.

The *CRC Handbook of Processing and Utilization in Agriculture* (8) contains an extended discussion of the dairy industry, for additional detail.

I. DAIRY INDUSTRY

Some of the largest and best-known food companies have their roots in the dairy industry, though most have diversified widely over the years. For example, Kraft, Beatrice, Carnation, Pet, and Borden were all originally dairy companies and most still produce some dairy products. Consolidations, mergers, acquisitions, and divestitures have rearranged operations among these companies as they have with many others. For instance, Borden purchased most of Beatrice's dairy operations in 1986 and Nestle, itself a huge dairy processor internationally, owns Carnation.

In contrast to some traditional areas of the food business, dairy products have seen steadily increasing production. Milk production increased about twenty-fold between 1950 and 1980 (Figure 1). Of course, not all of this increased production was consumed, leading to the well-known problem of surpluses in the form of nonfat dry milk, butter, and cheese. However, consumption did increase for such products as yogurt, ice cream and low-fat milk. Understanding the manufacture of some of the major dairy products is thus a useful and relevant exercise.

II. PROCESS OVERVIEW

Figure 2 shows the general process flow for several dairy products. Raw whole milk is received and stored. Provisions must be made to receive raw milk seven days a week, even if the dairy only operates five or six, because the dairy cows produce every day and on-farm storage is usually limited. (For some purposes, milk can be concentrated on the farm, permitting easier storage and reducing transportation costs.)

Raw milk is naturally contaminated with microorganisms which can grow to cause spoilage and, in some cases, disease. The design and operation of a dairy is largely directed at preventing these undesirable events while maximizing the efficient production of high-quality products. Manipulation of temperature is one of the major tools used to protect dairy products. Low temperatures, above freezing but below 40°F, reduce but do not eliminate the rate of growth of most bacteria. High temperatures, from 145°F to 300°F, can kill many bacteria if applied for sufficient lengths of time (30 minutes at 145°F, or 16 seconds at 160°F). High temperatures also affect the flavor of dairy products, generally unfavorably.

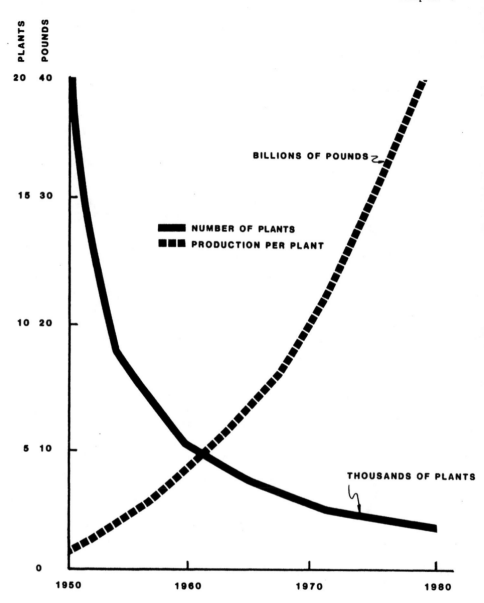

FIGURE 1 Milk industry trends.

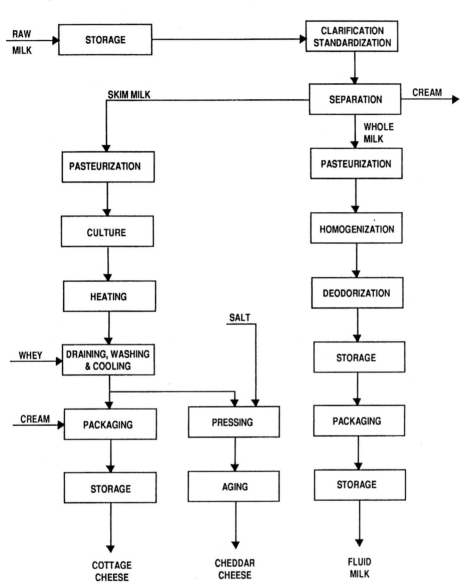

FIGURE 2 Block flow diagram for dairy products.

Since milk contains both proteins and sugars, it is vulnerable to the Maillard browning reaction discussed in Chapter 3. In addition, heating can induce carmelization of sugars and other off-flavor-forming reactions. Thus, dairy processing is constantly navigating between the constraints of preservation and flavor. In this respect it is similar to many other parts of the food processing business.

The other major theme in dairy processing is the manipulation of composition to manufacture various products. The gross composition of milk is shown in Table 1.

It is obvious that milk is relatively dilute. It also is a multiphase emulsion of fat partially stabilized by proteins. There are many proteins in milk, including about 14 caseins, lactoglobulin, and lactalbumin (see Chapter 3 regarding proteins). While their amino acid compositions vary, collectively they are very nutritious for humans. Functionally, they are characterized by their solubility behavior and other properties, such as their tendency to form solid curds, stabilize foams, and form complexes with ions or fats.

The fat in milk is relatively saturated or hard by comparison to vegetable oils. This contributes to the familiar properties of butter, a semisolid at room temperature.

The fat in milk has a lower density than water, which provides the basis for a physical separation. Under the influence of gravity, the fat will rise to top of milk, forming a separate layer known as cream, which can contain from 20 to 80% butterfat. The rate of this separation can be increased and made continuous by imposing an artificially high gravitational field in a centrifuge. This separation is usually made as completely as possible and then other products are made by reblending. The aqueous phase is known as skim milk and contains little fat, but most of the soluble protein and sugar (lactose).

TABLE 1   Gross Composition of Milk (Average)

Substance	Percentage
Water	87.8
Fat	3.7
Protein	3.2
Lactose	4.6
Ash	0.7

The fat phase or cream may be further processed to make butter, used directly as cream, blended with other streams to make Half and Half, added to cheese, or used to make ice cream. To make various fluid milk products, only a fraction of the cream may be skimmed in the separator. Subsequent steps include pasteurization, homogenization, deodorization, flavoring or enrichment, storage, packaging and distribution. Some of these process steps are almost unique to dairy processing and so deserve some further explanation.

Pasteurization is a carefully controlled heat treatment imparting a combination of time and temperature designed to destroy health-threatening microorganisms without seriously affecting the flavor of milk, cream, or other fluid. Because the degree of treatment is just sufficient to eliminate the relatively heat-sensitive pathogenic (disease-causing) organisms (tuberculosis, for instance), certain other bacteria can survive. Thus, pasteurization is not sterilization. While originally developed in the milk industry, the concept of pasteurization has subsequently been applied in other areas of food processing, such as drinks, fruits and vegetables, and meats. Pasteurized products eventually can spoil even when properly stored and packaged because the surviving microorganisms will slowly grow and produce objectional flavors, such as acids. As will be seen, certain dairy products depend on the deliberate growth of desirable microorganisms. Pasteurization will be discussed in greater detail later to illustrate certain mathematical concepts of microbiology.

Homogenization is a physical process for reducing the particle size of the fat dispersion so as to stabilize the emulsion. Once a fat globule has a diameter below 1 micron, the force of gravity due to the density difference is compensated by the frictional drag force and so cream will not spontaneously separate. The size reduction is typically achieved by subjecting a fluid mixture to high shear by forcing it to flow through a small orifice under high pressure drop. Machines differ in the construction of the orifice and the means of adjustment. Some homogenizers use two stages in sequence to eliminate reaggregation that can occur after one stage. Wear can be high in homogenizer valves because of the high velocities and, probably, cavitation that can occur because of the high pressure drop. As with pasteurization, homogenization has been applied outside the dairy industry, to produce such products as salad dressings and sauces, but its largest application is still milk and other dairy fluids.

Deodorization is a simple process to remove objectionable flavors and aromas from milk by heating, exposing to vacuum, and cooling. The flavors typically arise from certain feeds eaten by dairy cattle, such as wild onion, wild garlic, or other odiferous herbs or weeds. Fortunately, these are relatively volatile compounds. Where herds receive a controlled diet all year, flavor changes are less likely.

Processed milk is stored in bulk refrigerated tanks until it can be packaged. Common packages are plastic gallons and half gallons, paperboard gallons, half gallons, quarts, and pints, and, for food service, plastic bags holding several gallons. Relatively few glass containers are used today, though they once were quite common. The plastic containers are commonly blow-molded on site, using automated machines and polyethylene resin. Paperboard containers are formed by heat sealing plastic-coated roll stock or preformed blanks.

After packaging, milk products are stored briefly in coolers, but because they have limited shelf life, they are distributed quickly. Frequently, pull dates are printed on retail milk packages so consumers are assured of fresh product. The goal is to provide 15–20 days shelf life under refrigerated conditions, that is, no objectionable flavor change perceptible over that period. Returned fluid milk can be used in other dairy products, including cheese, ice cream, and dried milk.

III. CHEESE

Cheese is an ancient food discovered empirically as a means to preserve and concentrate the food value of perishable milk. This probably occurred when milk stored in containers made from an animal's stomach was found to have formed a solid. The residual rennin (a digestive enzyme) would precipitate milk protein. There are many varieties of cheese valued for their distinctive flavors and textures which are developed by specific processing techniques and microbial cultures. Cream, cottage, and cheddar cheese are popular in the United States.

All cheeses begin by precipitating most of the protein from whole or skimmed milk. This precipitation is accomplished by addition of rennin, an enzyme found in the stomachs of calves (slaughtered for veal), or by manipulation of pH (see Chapter 3). If the acid content of milk is increased, the suspended and soluble proteins become insoluble and precipitate. The acid can be increased by addition or by permitting certain bacteria, which convert lactose to lactic acid, to grow. Since microbial growth is desirable in this case, the milk is maintained at a warm temperature in large tanks. The protein precipitate is initially relatively fine, but by gentle heating and agitation, the particles aggregate into "curds" which permit easier separation from the soluble phase and contribute to ultimate texture. If fat is present, as in whole or partially skimmed milk, it is occluded by the precipitating protein and found in the curd.

Washed curds blended with pasteurized cream in various proportions make the familiar cottage cheese, which is sold fresh and refrigerated. If the curds are pressed to increase their solids content and subsequently aged, then harder cheeses are produced. The drained curds from whole milk are folded and pressed repeatedly to develop the characteristic texture of cheddar cheese; the process is called cheddaring. More or less salt and various microorganisms are added to develop distinctive flavors, color, and texture in specific cheeses. Aged cheeses, with their higher solids content and low water activity, have relatively long shelf lives.

The distinctive flavors of such cheeses as Swiss, Cheddar, Blue, and Brie are achieved by inoculating with specific microorganisms and holding molded shapes of curd at relatively warm temperatures and high relative humidities. Proteolytic and lipolytic enzymes modify the protein and fat present to develop complex flavors. Slowly growing yeast generate the characteristic bubbles in certain cheeses, while special molds contribute to the color, flavor, and texture of Blue, Roquefort, and Brie.

It is interesting to consider the price paid for this shelf life extension of milk. Cheddar cheese has less than 39% water and is usually made from whole milk. This means that the solids in cheddar, 61%, come from the 3.7% fat and some fraction of the protein. The lactose and ash, being soluble, do not precipitate. About 80 to 85% of the protein (mostly the casein) and all the fat precipitates. By calculation, then, it takes about 10 pounds of milk to make 1 pound of cheese: 3.7 + 0.85 (3.2) = 0.61 X, where X is pounds of cheese from 100 pounds of milk, X = 10.5. By most standards, this is a relatively low yield. Further, it raises the question of what becomes of the remaining 90 pounds of milk?

IV. BY-PRODUCTS FROM CHEESE MAKING

The fluid remaining after cheese curds are separated is known as whey. It has been a serious disposal and utilization problem for many years. Because it contains the soluble sugar, ash, and protein, it is not pure water, and in fact, is a fairly strong waste stream in terms of biological oxygen demand (BOD). Measured compositions for typical wheys are given in Table 2.

Sweet whey results from precipitation with rennin in making ripened cheeses. Bacterial cultures are used primarily for flavor in such cheeses as cheddar, Swiss, and Muenster. Acid whey results from precipitation by acid-producing cultures to make such cheeses as cottage and baker's.

Various uses have been proposed for whey, but each idea poses problems. As a waste, it is relatively strong and fairly difficult to treat biologically because lactose, the major carbon source, is hard for many organisms to digest. (It is also hard for some

TABLE 2 Typical Composition of Sweet and Acid Whey

Component	Sweet	Acid
Water	93.5%	94.8%
Protein	0.8	0.6
Lactose	4.9	4.3
Lactic acid	1–2	7–8
Ash	0.6	0.5

humans to digest, which is why certain cultures rely more heavily on cheeses and other processed dairy foods in which lactose is reduced as compared with whole milk.) Direct discharge to receiving waters is simply not legal in the United States any longer. Further, recovering some economic value from whey is certainly appealing when it represents such a large fraction of the purchased raw material.

The solids in whey, dilute though they are, contain nutritive value for animal feed or human food. The nutritive value of whey proteins is less than that of whole milk because the proteins are fractionated during the cheese making process (see Chapter 3 regarding protein solubility). For either use, the solids should be concentrated somehow for convenience and ease of handling. Several methods have been proposed and will be considered in some detail later.

Another alternative use is fermentation to such materials as alcohol, yeast, or methane. Again, the relative indigestibility of lactose is a potential obstacle, but there are solutions. Organisms can be selected which thrive on lactose, enzymes which aid its digestion can be added, or whey can be supplemented with other materials to make a more complete fermentation medium. When oil prices are high, producing fuel alcohol from waste whey is attractive.

In any attempt to use whey, there are additional concerns. As a large-volume, low-value material, it usually is treated as a nonfood, which increases the risk of contamination by undesirable microorganisms. This not only reduces its value further, but it can provide a source of contamination to food products, especially inoculated cheese in which a relatively pure culture is desired. Wild yeasts, bacteria, and molds typically can outgrow selected pure cultures and thus spoil a batch of cheese very quickly. Thus, handling and storage of whey demands the same standards of sanitation as are required for milk, if it is not to become a hazard to the main process. An additional obstacle to the economic use of whey is the relatively low volume found in any one plant. Since about 90% of whey is water, it requires a large amount to produce significantly interesting amounts of concentrate or by-products. Nonetheless, driven by legal pressures concerning waste disposal and the development of effective new processes, processing of whey is increasing.

V. MEMBRANE PROCESSING

Concentration of whey by evaporation of water is relatively expensive because of its energy consumption (500–1,000 BTU/pound of water, depending on the efficiency of the evaporator). An interesting alternative is membrane separation, ultrafiltration or reverse osmosis, depending on the membrane used, pressure and degree of separation required.

In contrast to conventional filtration, which relies on gross differences in physical dimension to separate particles from liquid, certain polymeric or inorganic membranes are capable of retaining molecules or very small particles while transporting solvent molecules and low-molecular-weight materials. There are many membrane materials and configurations of equipment which differ in cost, selectivity, ease of operation, and stability. Usually, experimentation on specific equipment and feed material is necessary for accurate design. For illustration here, data available in the literature is used for representative materials.

Example: Ultrafiltration of Sweet Whey

Assuming the whey composition given in Table 2, compositions have been calculated for various amounts of water removal. Permeable materials (lactose, lactic acid, and ash) maintain about the same concentration on both sides of the membrane, while protein is completely retained on the high pressure side. See the schematic flow diagram (Figure 3). Additional information is required to complete the material balance, namely, the total solids found at a given water removal. Timmins (6) provides such data (see Table 3).

There is an apparent inconsistency in the data (not at all an unusual event), which should be reconciled before proceeding. Most likely, the lactose concentration does not actually remain constant, but no obvious adjustment cures the inconsistency. (The inconsistency arises from the assumption that all the protein is retained and the correlation of total solids with water removal: 0.8/2.46 = 32.5% protein compared with 24% total solids, obviously impossible.)

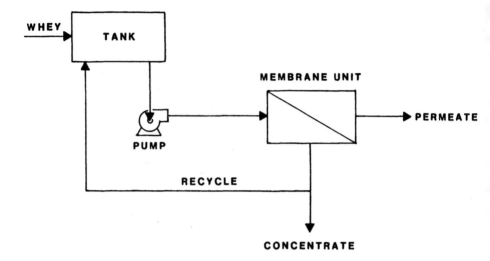

FIGURE 3   Schematic flow diagram membrane processing of whey.

One approach is to recalculate water removal, requiring all the other
compositions to remain as assumed. That is, protein becomes a tie
component, since it is retained, total flow is calculated as 0.8/frac-
tion of protein (%/100), and water is calculated by difference. Water removal is then water content divided by 93.5. Results are
shown in Table 4. The water removals are close enough to those
originally assumed that the revised table can be used to compute
a material balance. It is worth noting that there is some danger
in using a minor component as a tie in this exercise, because errors can multiply quickly, but there is not much choice in the absence of more precise data. It is essential that data used for design be consistent, even if some manipulation must be performed
to achieve that state.

    There is a relationship between flux and protein concentration.
Temperature and velocity are also important variables. Before proceeding, it is useful to explore the significance of flux and consider the impact of other design variables.

    The cost of an ultrafiltration system depends primarily upon the
amount of membrane required for a given flow rate. This, in turn,
depends on the permeation rate, or rate of flow through the membrane. Permeation rate is dependent on pressure drop and tan-
gential flow velocity as well as viscosity. Viscosity is affected
by temperature and solids concentration. A phenomenon called

TABLE 3 Material Balance on Ultrafiltration

Component	Weight percent of component by percentage of water removed				
	0	50	75	95	98
Protein	0.8	1.8	3.8	11.3	18.3
Lactose	4.9	4.9	4.9	4.9	4.9
Lactic acid	0.2	0.2	0.2	0.2	0.2
Ash	0.6	0.6	0.6	0.6	0.6
Water	93.5	92.5	90.5	83	76
Total solids	6.5	7.5	9.5	17	24
Total pounds	100	50.54	25.83	5.63	2.46

Source: Data from Ref. 6.

concentration polarization can reduce the influence of pressure drop. As rejected solutes (protein in this case) build up against the membrane, like a filter cake in conventional filtration, the continued transport of solvent (water and dissolved low-molecular-weight materials) depends on diffusion across the high-protein-concentration region. Such diffusion is not influenced by pressure. This behavior has led to ingenious designs to enhance mass transfer by using narrow channels, turbulence promoters, and high velocities. For each solution there are penalties: narrow channels can easily plub with suspended solids, as can turbulence promoters (such as screens), and high velocities require increased pumping. High recycle rates, to achieve high velocities, may create sanitation problems in sensitive foods by increasing residence times.

TABLE 4 Reconciliation of Water Removal

Total stream, lb	44.44	21.05	7.08	4.37
Water, lb	41.11	19.05	5.88	3.32
Removal, %	56	79.6	93.7	96.4

Since ultrafiltration units are inherently modular in construction, that is, they consist of multiple identical units, the example can be worked on the basis of some nominal flow, such as 1,000 gallons, with scale-up to a particular case being straightforward. Volumes of about 50,000 gallons per day might be typical for a medium-sized cheese plant; large plants could generate ten times as much. If a flux of 10 gallons per square foot per day can be maintained, 5,000 square feet of membrane are required. This is a large device!

As a rough approximation, membrane-processing systems cost about $10 per square foot of membrane, more or less depending on details of construction. Thus, a unit to treat 50,000 gallons per day might cost about $50,000. For comparison, a waste treatment system to treat whey might cost about $2–3 per daily gallon treated or $100,000–150,000.

Three configurations are generally available for membrane devices: hollow fibers (typically 1 mm diameter, 100 cm long, 10,000 per module), spiral wound or flat plate (typically 100 sq ft in a module), and tubular (typically 0.25 inches in diameter and about 6 feet long). The hollow fibers have the highest density and the tubes, the lowest. Costs depend somewhat on density (square feet of membrane per cubic feet of device), so hollow fibers are economical when they are applicable; however, they are hard to clean.

Ignoring for the moment the specific configuration, we can consider several process flows. One choice is once-through flow, in which velocity decreases down the length of the membrane while concentration increases. For high water removal, the volume decreases significantly; flux declines, viscosity increases, and pressure drop increases. In practice, it is more common to recycle retentate so as to increase velocity across the membrane and buffer the impact of volume loss. One consequence is that the flux is more constant at a value corresponding to the product concentration. To obtain the benefits of high flux rates at low concentrations, it is possible to stage ultrafiltration units in sequence, each with its own recycle stream. This somewhat complicates the process but may reduce cost. Rudd et al. (5) explore some process alternatives.

VI. ICE CREAM

One of the most popular dairy products is ice cream in its many variations. The manufacture is relatively simple, but there are some interesting issues. Ice cream is a frozen mixture of milk solids, butterfat, sweetener, water, and air with appropriate flavoring and, in some cases, functional additives. It is prepared from cream, condensed or dry milk, sugar or corn syrup, water, and other materials. The liquid mixture is pasteurized and heat-sensitive components are added. Then it is frozen to a thick suspension, particulate components (such as fruit or nuts) are added, the mixture is filled into containers or, in the case of novelties, shaped in molds, and finally, the containers are further frozen to "harden" the product. Figure 4 is a schematic flow diagram. Bartholomai (1) says a small plant making about 1 million gallons per year would cost about $3.3 million to build and about $2 per gallon to run.

To improve texture, air is incorporated during the freezing process; control of the amount of air is important. Since ice cream is sold by volume, and not by weight, there is an understandable temptation to incorporate as much air as possible, since it is less expensive than butterfat, milk solids and sugar. Legal restrictions specify the minimum weight of a gallon, which translates to allowable "overrun," expressed as a percent increase in volume for a given amount of mix. Normally, the requirement is 4.6 pounds per gallon, which equates to 100% overrun for a 9.2 pound per gallon mix. Certain premium ice creams pride themselves on lower overruns, that is, higher density products (also making them higher in caloric content). This can have an interesting practical impact on equipment design.

Ice cream freezers are quoted in nominal gallons per hour, assuming 100% overrun, for example, 300 gallons per hour. If a product is to have only 20% overrun, the volumetric capacity of the freezer will be reduced by about 40%: 300 × 1.2/2.0 = 180. This is because the heat removal capacity of the freezer is fixed and depends primarily on the mass throughput rate.

After filling, containers or shaped novelties must be further frozen. It is important that this be done as rapidly as possible so that water freezes in the form of small ice crystals. If large crystals form, the ice cream has an unpleasant "sandy" texture. Ice crystallization is a mass transfer phenomenon controlled by heat transfer; if heat is removed rapidly, water solidifies on many nuclei and is immobilized. If heat is removed slowly, water has the opportunity to diffuse farther and contribute to the growth of larger crystals. Heat transfer within the ice cream is almost entirely by conduction, since the viscosity is so high and there is normally no agitation. Thus the only way to accelerate freezing is to reduce external heat transfer resistance. This is done by using the lowest practical temperatures (−40°F), high air velocities, and direct contact with cold plates (when shapes permit).

Because the hardening time is controlled by internal heat transfer, the modified Plank's equation discussed in Chapter 4 can be used to estimate freezing times. Ice creams contain about 60% water and can have freezing point depressions of over 50°F. That

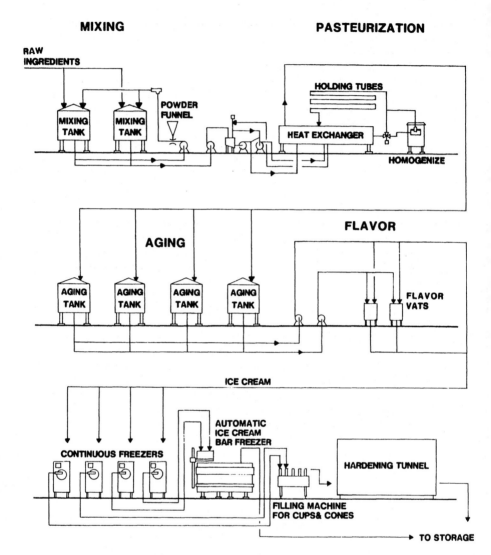

FIGURE 4   Ice cream plant flow schematic.

is, −20°F, about 85% of the water present is frozen. A common target for hardening is a core temperature of 0°F.

Ice cream from a continuous freezer is usually around 22°F, where about 40% of the water is frozen. If the outlet temperature is allowed to drop even a few degrees, to about 20°F, the slurry becomes quite viscous and may overload the motors driving the paddle on the freezer.

The high viscosity (because of the high solids content) of ice cream mix requires larger pumps, larger diameter tubing, and more vigorous cleaning practices than are found in milk processes. The fact that certain ingredients do not receive any heat treatment (frozen strawberries, for instance) increases the opportunity for bacterial contamination. The low temperatures at which ice cream must be stored greatly reduces the opportunity for bacterial growth, but ice cream that is abused (melted and refrozen, for instance) could pose a potential health hazard. Fortunately, such products would also become unappealing, because the texture would be dramatically changed (large ice crystals would be present), and this would serve as a warning that the product had been mistreated. Ice cream is not generally considered a source of health hazards, but its manufacture still requires care and sanitary design.

One recently recognized risk to ice cream is *Listeria monocytogenes*, a pathogen which grows at low temperatures and which has been found in dairies. This microorganism can cause a disease which has been fatal to unborn children and to immune-suppressed people. Past outbreaks have been traced to Mexican-style soft cheese and to a processed meat. It has been isolated from frozen novelties, leading to product recalls and plant closures.

Rigorous sanitation is the only prevention of contamination by *Listeria*. In particular, careful use of hoses, especially around drains, is suggested because the organism can be spread in the aerosol created by spraying water around. For the same reason, hub drains are suggested instead of trench drains in most situations. In general, a dry floor is more desirable than a wet one, to reduce the risk of contamination, and also for worker safety and comfort.

VII. CONCEPTS OF THERMOBACTERIOLOGY

Previously, pasteurization was described as a preservation technique which relies on increased temperature to reduce bacterial populations. More intense heat treatments lead to greater reductions; when bacterial or spore populations are totally eliminated, the product is considered sterile. Heat treatments also inactivate enzymes and other chemical processes while accelerating certain changes. (In milk, the enzyme alkaline phosphatase has thermal processing characteristics similar to those of pathogenic organisms, thus analysis for the enzyme serves as an indicator of proper processing.) Defining a heat treatment process for food is a complex affair. However, the essentials can be described using the concepts of reaction kinetics.

Bacterial, spore, or enzyme population decline with time at a constant, damaging temperature is a first-order process (Figure 5). (The temperature must exceed some threshold to be damaging; at lower temperatures, populations may increase due to growth or remain constant. For most organisms of interest, temperatures must exceed 100°F to be damaging.) A first-order process has the mathematical solution: $\log (n/N) = -kt$, where n is the population or concentration at time, t; N is the initial value at time $= 0$; and k is a rate constant. Since such an equation gives a straight line on semilogarithmic paper, the rate constant is often expressed as its reciprocal D, the decimal reduction time. This is defined as the time to reduce a population by one log cycle, that is, a factor of ten. Multiples of D are often used to describe thermal processes, for example 12 D, or reduction by a factor of 10 exp -12.

If the units measured are individuals (cells or spores per container, for example) and if the initial value is a value less than the achieved reduction, 10 exp 8, for instance, than a thermal process calculation produces the apparent paradox of a population value

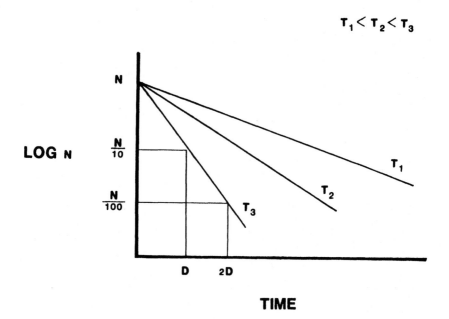

FIGURE 5   Temperature effect on microbial population.

less than one. This is interpreted as a probability that a cell or spore will survive the treatment. Thus, in a common practice for canning, where a 12 D process applied to an assumed initial value of 10 exp 8 gives a probability of 10 exp −4, the expectation is that one container in ten thousand might spoil. In practice, the canning industry does even better, but the fact that some containers do spoil even after proper processing is interpreted as confirming the mathematical model.

If experiments on target microorganisms, enzymes, or spores are performed at different temperatures, different values of decimal reduction times (rate constants) are found. This is also true for chemical reactions, and is expressed in the Arrhenius equation: log k = A −E/RT, where k is the rate constant, A is a constant, E is called the activation energy, R is the gas constant (correcting for units), and T is absolute temperature.

The temperature effect on biological rate equations has also been expressed in other forms, chosen for convenience or as useful approximations. The most common is z, defined as the temperature change required for a tenfold change in decimal reduction time. Values for these parameters measured for several organisms are in Table 5, taken from Jackson and Lamb (3), who in turn compiled these from several sources.

In this table, F is the time in minutes to achieve a 12 D process with a pure culture in a defined medium, at a specified constant

TABLE 5 Data for Thermal Death Rates of Microorganisms

Species	E (kcal/ mol)	A (kcal/ mol)	D (min at 250°F)	F (min at 250°F)	z (°F)
Bacillus stearother-mophilus FS1518	68.7	38.13	3	36	18
Clostridium sporo-genes PA3679	68.7	38.22	1.3	15.6	18
Bacillus subtilis FS5230	68.7	37.98	0.75	9	18
Clostridium botulinum	69.24	39.45	0.21	2.45	18

temperature, in this case, 250°F. In practice, using real foods in
real systems, longer process target times are specified. Further,
the ideal of constant temperature is difficult to achieve in practice
(though it can be approached in flowing systems) and so processes
with temperature varying over time are numerically integrated, using
the temperature corrections described, to compute the contribution
to lethality of each time increment.

The integrated lethality F is defined as:

$$F = \int_{t_1}^{t_2} 10^{-(250 - T)/z} \, dt$$

where t_1 and t_2 are the time of start and end of the process, T is
temperature at any time, and z is the temperature correction.

Processes having different time-temperature profiles are considered equivalent if their computed integrated lethalities are equal.
Some representative commercial processes are shown in Table 6.

Ingenious techniques have been derived to simplify this calculation, but with modern computers, these are less necessary. Most
food science texts elaborate on the subject; the purpose here is
to introduce the basic concepts. For example, see Charm (2) and
Richards (4).

Other reactions of interest to food processing are also affected
by heat. These include vitamin destruction, color development

TABLE 6 F Values for Commercial
Sterility of Some Foods

Food	Lethality (min at 250°F)
Carrots in brine	3—4
Cream soups	4—5
Evaporated milk	5
Peas in brine	6
Meats in gravy	12—15
Pet foods	15—18

Source: From Ref. 3.

(browning or Maillard reaction), enzyme inactivation, flavor development (especially carmelization), and texture change. With few exceptions, the activation energy of vitamin destruction (and other harmful reactions) are smaller than the values for cell and spore destruction. [This means the lines of log k versus reciprocal absolute temperature for vitamin destruction are more shallow than those for spore destruction (Figure 6)]. From a process standpoint, the consequence is to favor high-temperature, short-time processes to achieve desired sterility with minimum destruction of nutrients. Nutrients behave like less easily quantified properties, such as flavor and color in the sense that, normally, high temperatures and short time produce higher quality products than other processes. For example, milk pasteurized at 160°F for 30 minutes will taste more "cooked" than that processed at 180°F for 30 seconds. When this logic is extended further, "ultrahigh-temperature" (UHT) processes result, using temperatures near 300°F for fractions of a second. In the case of milk, a shelf-stable (essentially sterile) product results with quite acceptable flavor. However, inactivation of enzymes is incomplete because their activation energies are very low and process conditions adequate for microorganism destruction do not completely inactivate enzymes, such as lipase, which could cause undesirable flavor development over time. Thus UHT processes must be increased slightly beyond the time determined for sterility in order to improve shelf life. The flavor is slightly poorer as a result.

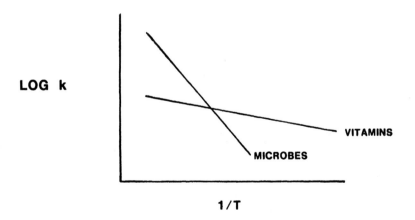

FIGURE 6 Comparison of rates of destruction of vitamins and of microbes.

High-temperature short-time pasteurizers (HTST) use plate heat exchangers with a high degree of regeneration. Ultrahigh-temperature processes may use direct steam injection or indirect heating in tubes; if direct steam injection is used, the treated fluid is cooled very rapidly by exposure to vacuum, which also removes the water added as steam (and some volatile flavor). The taste of UHT milk changes slowly during storage at refrigerated temperatures and more rapidly at higher temperature, because the Maillard reaction can occur at a rate that increases with increasing temperature.

One motivation for extended shelf-life dairy products is to minimize process changeovers. Specialty products, such as coffee creamer, chocolate milk, and drink mixes are inconvenient to make every day because their volumes are relatively low and they may be incompatible with other products, necessitating cleanup between short runs. Further, the fact that many specialties are flavored makes them less sensitive to the taste changes that are induced by higher temperature processing. Thus the most common applications of UHT processes are for such specialties.

Sterilization or pasteurization processes are also affected by pH and water activity. For example, *C. botulinum* does not produce toxin at a pH below 4.6. Therefore, foods which are naturally acid or which have been acidified can be subjected to milder heat treatments, sometimes no higher than 212°F.

Low water activity also inhibits microbial growth, so that foods high in soluble solids usually do not require rigorous heat treatment.

VIII. CLEAN IN PLACE

Before the development of automated cleaning techniques, the common practice in dairies was to manually disassemble all the process equipment and piping so it could be washed and sanitized by hand. This was labor intensive, subject to human error, and so time-consuming that it limited the practical size of dairies. The concept of cleaning equipment and piping without disassembly permitted the construction of larger and more efficient dairies with less labor intensity.

The essential features of a clean in place (CIP) system are as follows (Figure 7):

Welded piping system (less expensive than threaded fittings and more reliable)

Welded CIP supply and return system (additional piping for cleaning solutions)

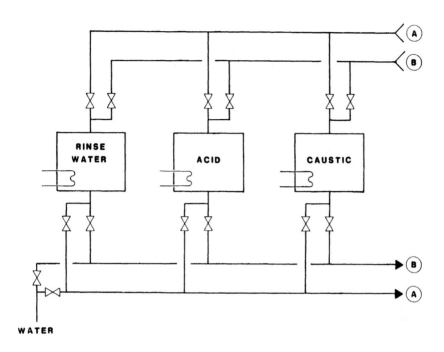

FIGURE 7   Schematic diagram of centralized CIP system.

Permanent spray devices in tanks (spray balls or nozzles designed
   to ensure complete coverage of all surfaces)
Automatically controlled CIP circulating units (multitank, valve and
   pump systems which circulate cleaning, rinse, and sanitizing materials for specified times through specified circuits and control
   the addition of cleaning agents (originally controlled by electro-
   mechanical devices, but now use electronics)
Air-operated valves (in contrast to hand operated plug valves)
Centralized control (one or more systems for an entire plant

   The soils encountered in dairies are primarily derived from milk,
of course, and so are fatty or proteinaceous.   Under heating, some
of the calcium in milk can contribute to proteinaceous deposits called
milk stone.   These soils are vulnerable to solubilization with caustic
and, if necessary, acid.   To reduce costs and to minimize discharge
of wastes, cleaning chemicals are reused in CIP systems, with chemical addition under automatic control to make up for losses.   Caustic
concentration, for example, is commonly maintained at 1.5%.

Calculation of proper circulation rates is important since soil re-moval is difficult to confirm in a closed piping system. Practice in the United States is to design for a minimum velocity of 5 feet per second. In Europe, it is recognized that the critical factor is suf-ficient turbulence, which is characterized by the Reynolds number (mass flow rate times diameter divided by viscosity). A velocity of 5 feet/second in 1-1/2 inch tubing (for water) corresponds to a Reynolds number of about 80,000. In principle, it should be the dimensionless value which remains constant if the diameter of tubing changes; in American practice, the velocity is kept constant. One consequence is higher pumping costs than necessary.

The concepts of CIP are potentially applicable to other food proc-esses, but experiment is necessary to establish effective cleaning conditions for the specific soils encountered. Caustic is still a good, general-purpose agent, but concentrations may need adjustment to affect starch-based soil, as in a cereal plant. Settling and filtra-tion may be necessary to treat recovered cleaning agents and to concentrate removed soil for disposal.

REFERENCES

1. Bartholomai, A., ed., *Food Factories*, VCH, New York (1987).
2. Charm, S. E., *Fundamentals of Food Engineering*, AVI, Westport, Connecticut (1971).
3. Jackson, A. T., and J. Lamb, *Calculations in Food and Chemical Engineering*, Macmillan Press, London (1981).
4. Richards, J. W., *Introduction to Industrial Sterilization*, Aca-demic Press, London (1968).
5. Rudd, D. F., G. J. Powers, and J. J. Siirola, *Process Syn-thesis*, Prentice-Hall, New York (1973).
6. Timmins, R. S., Whey and skim milk, *Freeze Drying and Ad-vanced Food Technology*, edited by S. A. Goldblith, L. Ray, and W. W. Rothmayr, Academic Press, London (1975).
7. U.S. Department of Commerce. *Prepared Foods New Products Annual*, pp. 55–58 (1988).
8. Wolf, I. A., ed., *CRC Handbook on Processing and Utilization in Agriculture, Volume 1*, CRC Press, Boca Raton (1982).

8

Orange Juice Case Study

This case was chosen to illustrate a number of interesting characteristics of the food industry that are a little bit unique and because it is an important category of the food business. The topic is orange juice, specifically frozen concentrated orange juice. Some of the characteristics arise from it being a seasonal crop, a tropical fruit harvested over a relatively long season. Oranges, and citrus in general, are harvested for about 6—9 months; nonetheless the fresh fruit is not literally available year round in the United States, and so it has to be processed and preserved. It is an important and popular source of nutrition, particularly as a natural souce of vitamin C.

The industry in the United States is a fairly recent development and it is still evolving and changing. The first frozen concentrated juice was made in 1945. Some of the largest companies in the food business are in the orange juice business. Beatrice owned (and recently sold to Seagrams) Tropicana, which is a very large processor, General Foods owns BirdsEye, Coca-Cola is involved with its Minute Maid brand, and Proctor & Gamble recently entered the business.

In 1988, over 146 million boxes (90 pounds each) of oranges were processed, down from as much as 218 million boxes in 1980 (5). At prices of $8.52 per box in 1988 and $6.15 in 1979, the raw material for frozen concentrated orange juice has a value of $1.3—1.8 billion per year. (Unit prices are $0.06—0.08 per pound of oranges.)

Florida produces over 90% of the oranges processed.

In 1987, frozen concentrated juice production was about 170 million gallons. Apparent yields, therefore, are about 1.3 gallons/box. These statistics are consistent with a 50% yield of juice, as we shall see later (1.3 gallons concentrate × 4 gallons juice per gallon concentrate × 8.5 lb per gallon = 44.2 lb juice per box divide 90 lb per box = 49%).

Figure 1 is a schematic diagram of an orange cut in half. The juice is contained in small sacs which must be broken to release the juice. Remnants of the sacs contribute pulp and a desirable mouth feel to orange juice.

What are some of the issues to consider in looking at preserving orange juice? First is consistency to the consumer. This means the same sweetness, same tartness, same color, and same characteristic flavor from container to container. With a natural product,

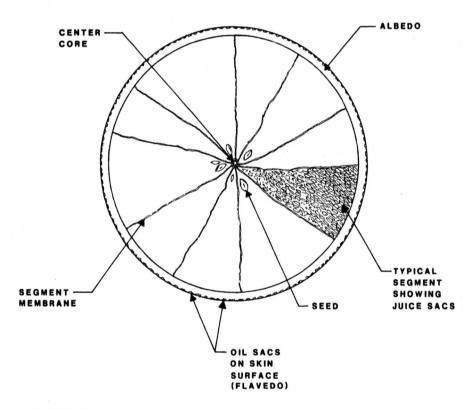

FIGURE 1   Schematic structure of an orange.

that is not easy to achieve. Oranges change in their sugar and acid content, flavor, and color with time of the year, with geographical area, varieties, and many other variables.

There are legal constraints on solids content and sugar and sugar-to-acid ratio by the state of Florida, and to a lesser extent by the federal government, for the protection of the industry.

The standards of identity specify that a 3 plus 1 concentrate (3 volumes of water to 1 of concentrate) shall yield a juice with not less than about 12% soluble orange solids. (The value was 11.8% in 1981 and the industry proposed a higher value of 12.3% in 1982.) Additional sweetener is allowed only if the label declares that it has been added.

The orange industry is probably the single most important economic factor in the state of Florida because about 95% of the Florida crop is processed, producing several billion dollars in sales each year.

A second factor is material handling. One of the things that is interesting about foods that chemical engineers never study, or study to a very limited extent, is handling solids. They learn how to pump fluids and compress gases, but rarely consider solids. Relatively few food products are pure fluid. The dairy industry starts out with fluid milk yet cheese requires solids handling. An interesting aspect of oranges and many other similar situations is not only handling solids, which are intrinsically a potential problem, but handling them in a non-steady-state fashion. A truckload of oranges from an orange grove weighs 40,000 pounds. It gets dumped in a very brief period of time, deluging the system with several tons of baseball-sized objects that are difficult to handle. This results in a sporadic process rather than a nice smooth steady state.

The same thing could also be said about a tomato processor which gets truckloads of tomatoes all at once and then nothing for awhile and then another truckload. Sporadic deliveries of solid is a unique problem to think about. It means one needs equipment and systems which are designed not for their steady state or their average capacity but for their instantaneous loading. A total of 10,000–40,000 pounds of oranges may be dumped in perhaps 15 minutes where the capacity of the plant may actually use that over a period of perhaps an hour.

The biological issue of enzymes and their control by some process arises in oranges and in some other fruits and many other foods. In this case, the goal is to prevent their natural behavior, specifically that of pectinesterase, which affects the appearance of juice.

Interestingly enough there are other cases where exactly the same enzymatic activity is encouraged instead of being discouraged. Specifically, orange juice is expected to be cloudy with suspended solids. The carbohydrate gum, pectin, helps maintain the suspension.

An enzyme, pectinesterase, attacks pectin, causing the juice to clarify. Some other juices are preferred to be clear (apple and cran-
berry, for instance). In these, enzymes may be added to promote
clarification.

Frozen concentrated orange juice is a concentrated product in
that the water is removed and solids are increased. The traditional
way has been by evaporating water, which is an energy-intensive
process. This is done on such an enormous scale in the orange
juice business that there has been some very sophisticated development in energy efficient concentration.

The flavor is popular, almost universally popular, very characteristic, very difficult to imitate and it is very sensitive. Orange
processing teaches some lessons that can be extended to other commodities in which heat-sensitive, volatile characteristic flavors are
significant.

Finally, juice represents perhaps 50% of the initial weight of the
raw material. That means that there is just as much nonjuice as
there is desired product, which creates another whole industry,
whose gross feed is exactly equal to the desired product. It is
not quite as bad as the cheese industry, where there was a 9 to
1 ratio, but still there are substantial quantities of a potential
waste material.

The dissolved solids of the juice contains a high percentage of
mixed sugars, about 76% of dissolved solids (1). The sugars are
about 50% monosaccharide but during storage, sucrose is inverted.
The juice contains a large quantity of organic acids, particularly
citric acid, and a small quantity of free amino acids. There is
some protein, some ash, in particular potassium-containing salts,
which are good from a nutritional standpoint, and a high quantity
of vitamin C and also vitamin A. Finally, there is a small quantity
of lipids, ether-extractable material, which are important constituents of the flavor. There are a variety of other chemical compounds. Some of these other chemical compounds are fascinating
in their own right and have become the basis of quite an interesting
small chemical specialty business. Table 1 summarizes orange juice
composition.

Figure 2 is an overall flow of orange usage showing the approximate fractions directed to various products. It shows that about
28% of the weight of all oranges grown is discarded as evaporated
water, 44% must be handled as by-products, and only 10% is de-
livered as concentrate.

Figure 3 is a flow sheet for orange juice production, prepared
as part of a study of different food processing industries (4).
Oranges are washed, because the surface of the orange can have
pesticides, dirt, sand, and insects adhering to it. A small amount
of surfactants may be added to the wash water to help remove

TABLE 1   Composition of Solids in Orange Juice

Component	Percentage
Sugars	76.0
Sucrose	50—60
Glucose	20—25
Fructose	20—25
Organic acid	9.6
Amino acids	5.4
Ash	3.2
Vitamins	2.5
Lipids	1.2
Other	2.1

contaminants. The oranges are graded by manual inspection to remove any that are bruised, moldy, cut, or unsound in any fashion.
They are sized by a clever mechanical system according to diameter.
The sizer consists of diverging cables on which the oranges roll.
When the orange reaches a point where the cables have diverged
beyond its diameter, the orange falls to a collecting conveyor or
bin. Fruit within the desired size range is collected by chutes.
   The same principle is applied to sizing other fruit, such as dates,
and vegetables, such as artichokes, where grading or subsequent
processing depends on size.
   They are almost perfectly spherical and are generally fairly uniform in any given orchard because of the way they are grown, but
it is important to have very uniform size because of the way they
are subsequently treated.
   The extractor is not a solvent extractor but a juice extractor.
There are three approaches:

1. Cut in half and ream (like at home).
2. Insert a porous tube and squeeze fruit.
3. Remove peel and crush fruit.

   Why not do something else? For example, why not just take
oranges and cut them up into small pieces and then squeeze that
mass? That is what is done to apples, for example.

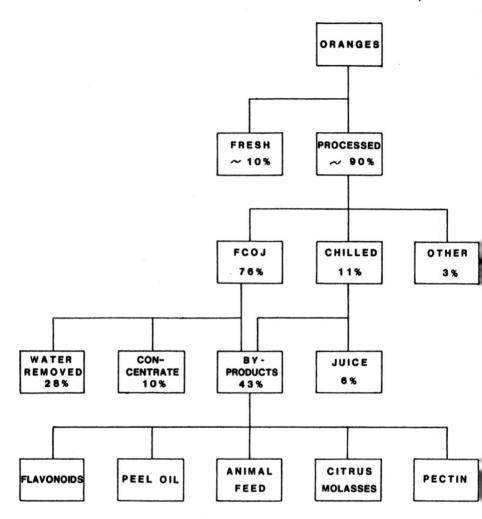

FIGURE 2   Overall flow of orange usage.

The skin is filled with oil, a very aromatic oil that is also rather bitter to the taste. The extraction devices that have been developed have been designed to exclude the peel oil from the juice. It turns out that if one were completely successful at doing that, the juice would not taste quite right. It needs a little bit of peel oil, but not too much. Furthermore, that oil is valuable as a flavor and

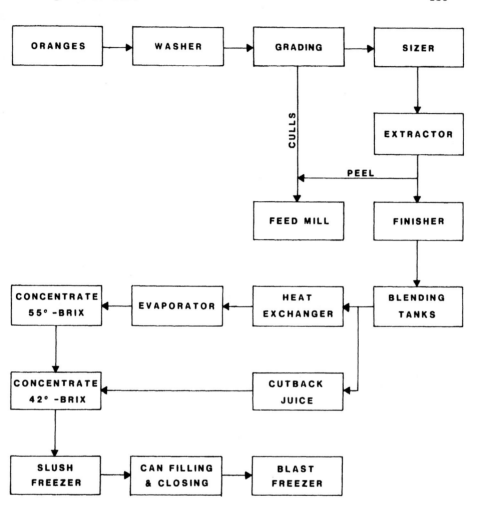

FIGURE 3   Orange juice production.

aroma ingredient in other products, so these complex, clumsy techniques for extraction are designed to keep separate two of the important parts of the orange.

The peel goes to a separate treatment and the juice then goes
through the rest of the process.

The next step is called finishing. This involves a scraped surface
screen which allows juice and small particles of pulp through while

retaining seeds and what the industry calls rag, the membrane and
the core. These are standard machines, which are also used for
tomato and other purees and pulps. Finishing is a size separation:
Big pieces are separated from fluid and small pieces. At this point,
the second attempt is made to standardize the quality.

When the oranges are received at the beginning, they are tested
and stored in separate bins by lot. When it comes time to process
them, oranges from different bins are mixed in order to standardize
the sugar, acid, and color. After the juice is produced, it is stored
in tanks and again tested so it can be blended in a further attempt
to achieve a uniform product. Thus, the raw material is blended
and the juice is blended in an effort to hit a certain target of total
solids, sugars, acids, and color.

The juice is heated to inactivate an enzyme called methyl pectin
esterase, which would decompose the naturally occurring pectin in
orange juice that helps keep orange juice cloudy. Fresh orange
juice initially is cloudy, but if allowed to stand enzyme modification
of the pectin will allow the pulp to settle to the bottom, leaving a
clear juice (serum) on the top. The flavor or, for that matter, the
color have not been lost, but it is unsightly. (This separation can
be used to advantage, as discussed later.)

Pasteurization of the orange juice, a delicate heat treatment, inactivates the enzyme. The juice is heated to temperatures of 145–
160°F for 5–30 seconds. This inactivates pectinesterase and also
reduces the microbial population. Some spoilage microorganisms can
survive in orange juice and concentrate, but few pathogens have
been found, fortunately. Offsetting the desire to pasteurize the
juice is the fact that the heat treatment has to be very carefully
controlled to preserve fresh flavor.

The juice is then concentrated in an evaporator. The heat exchanger serves as a preheater for the evaporator, but the evaporation is conducted under vacuum and at as low a temperature as can
be managed to prevent the development of cooked flavor.

In spite of all the precautions taken, the flavor in orange juice
is volatile and as water is removed a good portion of the flavor is
also removed. Reconstituted concentrated orange juice at this point
would be quite bland in flavor, and not characteristic of orange
juice. To compensate in part for that, a certain portion of fresh
unheated, untreated juice is added back to overconcentrated orange
juice to yield the concentrate of commercial practice, 42° Brix.
(Brix is a measure of the density of a sugar solution proportional
to the solids concentration. It can be correlated with refractive
index.) A 55% solid concentrate comes out of an evaporator and
is diluted with approximately 12–14% solids fresh juice to give 42%
solid concentrate, which is the conventional 3 to 1 concentrate. A
material balance shows that "cutback" juice is a substantial fraction

(about 30%) of the final product. The final product clearly has less flavor than fresh orange juice but more than pure concentrate does. While this process is traditional commercial practice, it is not the only approach.

After the concentrate is blended to 42°, it is frozen as a fluid by moving it through a chilled scraped surface heat exchanger, like an ice cream freezer, and then as a slush (slurry of ice and concentrate), it is deposited into containers and frozen solid. For certain purposes there are larger containers of concentrate for institutional or industrial purposes. Orange juice is used as an ingredient in some soft drinks, fruit juices, and some frozen products.

However, even with the addition of cut back, frozen concentrated orange juice is not confused with fresh orange juice. There is a desire to approach fresh flavor more closely. What are the characteristics of orange flavor? First, as many components are very potent, the concentrations in juice are very low, giving a very dilute mixture of very volatile compounds. They are volatile in spite of the fact that some of them actually have boiling points higher than that of water, because in dilute solution, they are highly nonideal, with high activity coefficients.

It is an established fact that in spite of their boiling points and molecular weights, the components of orange flavor disappear during evaporation. They are also sensitive to oxidation. What can be done? One approach, already discussed, is to add fresh juice which has all of the original flavor; about 30% of the final product is fresh juice when this is done. Another is to condense the vapor from the first stage of the evaporation and somehow distill the flavor away from the water. This yields a product called essence. It is not 100% of the flavor, but it is a lot of it. Both the Eastern and the Western labs of the Department of Agriculture have developed processes which are in commercial use to recover flavor essences from orange, apple, and some other fruits. That is a very neat chemical engineering solution, which yields a good, highly concentrated flavor essence which can be added back to the product and greatly improve the final flavor.

A small quantity of peel oil (0.025%), which can be recovered from the peel, can greatly enhance the flavor. Much more than this gives a noticeable bitter flavor.

Alternatively one can concentrate the juice in some way other than evaporation. One example is freeze concentration and another is selective membranes. Membranes permit a kind of filtration in which water is permitted to pass through or permeate and sugars or other larger molecular weight materials are not. Because of its low temperature, it does not volatilize the flavors and so membrane-concentrated orange juice would have a very good flavor.

As discussed later, membrane concentration of orange juice is difficult because of clogging of the membrane by suspended solids (6); however, a joint venture of DuPont and FMC is marketing such a system.

Freeze concentration is a process in which the water in a solution is turned to ice. Because ice is a crystal, it excludes anything else. If the ice can then be physically separated from concentrate, because of its low temperature, in principle, there should be a very high quality product. The energy requirement is less than that to boil water. One of the consequences is that the final product has enzymes present in small quantities and so juice will now separate if the enzymes are allowed to work. They have not been inactivated.

There is an interesting application of freeze concentration to orange juice.

Procter & Gamble has introduced a product called Citrus Hill, which is a frozen concentrated orange juice. For all practical purposes, frozen concentrated orange juice, because of common technology and because of regulations, is a commodity. There are brandname frozen concentrated orange juices that are marketed but it is hard to distinguish one product from another. At least that has been true until recently. Procter & Gamble have promoted Citrus Hill on the grounds that it is a measurably superior product.

In the literature there is at least one Procter & Gamble patent concerning orange juice (8). It is a product patent, although everything in it describes a freeze concentration process. The patent describes a modification of a commercially available freeze concentration process to produce what is claimed to be a unique frozen concentrated orange juice. That means that if they can devise another way to make the same product, they do not have to use the process that they described.

They describe a freeze concentration process manufactured by a company in the Netherlands called Grenco, which freezes a feed material using a scraped surface heat exchanger much like an ice cream freezer. This is not the only way to freeze, but one of several methods. One could use a shell and tube heat exchanger or other techniques. Direct contact, for example, with some kind of cooling agent like liquid nitrogen, would also work. The ice is separated from the slurry using a wash column or centrifuge. Because of the viscous, cold concentrate and small ice crystals, this separation process turns out to be very tricky. Usually, a little bit of ice is melted and used to wash the rest of the ice so that pure water and concentrate are discharged (Figure 4).

The cloud and pulp in orange juice pose problems in freezing so the patent describes separating the pulp from the serum or clear

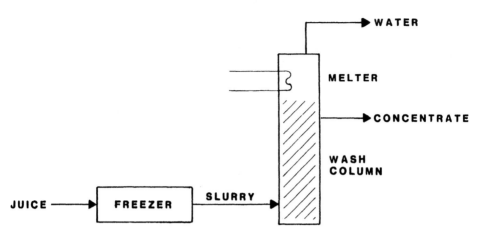

FIGURE 4   Freeze concentration.

juice by centrifugation concentrating the clear juice and then adding the pulp back. This is not the same thing as adding fresh
cut-back juice; this is a step around the physical problem of separating ice from cloudy concentrate by separating serum and pulp
and then recombining them. The mixture has stable cloud because
the pectin is retained in the serum, as is the pectinesterase. This
concept appears in an earlier patent, according to the Procter &
Gamble patent.

From thermodynamics, the heat of vaporization of water is seven
times the heat of fusion, so from a thermodynamic standpoint, freeze
concentration is less energy intensive than evaporation. However,
refrigeration on an energy basis is more expensive than heating and
multiple effect evaporation cuts this ratio even further. From an
energy standpoint, freeze concentration may have an edge, but it
is not 7 to 1.

The other traditional problem with freeze concentration is solids
loss. The solids in this concentrate are very concentrated, so small
fractions of entrainment in the ice are very costly, and very likely
because of the high viscosity of the concentrate which creates a film
on each ice particle. It has been suggested that freeze-concentrated
juice could be combined with evaporated juice at some advantage.
Freeze concentration is capital intensive (Table 2) and is being widely
investigated commercially. It is one of those great ideas that has
been looking for the right applications. The calculations shown in
Table 2 (7) were compared with those of several other authors and
found to be reasonable.

TABLE 2   Total Fixed Capital Costs ($000) 1981; Water-Removal Rate

Item	450 kg/hr	4,500 kg/hr	45,000 kg/hr
Triple effect evaporator	320	683	2,272
Mechanical vapor recompression	312	1,017	3,646
Reverse osmosis	433	1,877	11,550
Freeze concentration	1,720	4,700	30,900

Source: From Ref. 7.

From a chemical engineering standpoint the evaporator most often used for orange juice is interesting. It is called the TASTE Evaporator, which is an acronym for temperature-accelerated short-time evaporator, and it is one of the great success stories in the equipment business, because for all practical purposes today all orange juice evaporators are TASTE evaporators. Wine is also concentrated in a TASTE but it is almost exclusively associated with the juice concentration industry. It has replaced old, conventional high hold-up, high-temperature evaporators.

The essential ideas of this evaporator are, first, multiple effect, between four and seven effects. This means that the steam from one heat exchanger is used to heat the next stage of evaporation. One pound of steam can be used to evaporate as much as 3 or 4 pounds of water from the fresh juice. That obviously is an economical effect that is achieved by buying more metal for heat transfer surface. Today there are many effects, while a few years ago there might have been three or four effects. Energy has become more expensive and it makes more sense to buy capital to save on operating expenses.

The second characteristic of this machine is that it has a very short residence time, which it achieves by having a thin film rather than pools or large amounts of hold-up. It is a falling film evaporator with very short residence time and therefore very low thermal treatment. It uses relatively low temperatures—the highest is below the boiling point of water, about 210°F, and it goes down to about 90°F. This is achieved by maintaining a vacuum. It is designed to reach as high as 70% solids; since this would produce a very viscous material, in practice it is not generally done.

TASTE evaporators have been sized to handle a removal rate of several thousand pounds of water per hour and achieve about 2-1/2

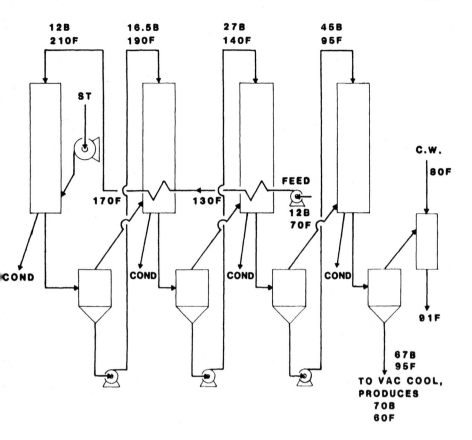

FIGURE 5 TASTE evaporator.

pounds of water per pound of steam. This is not equal to the number of effects because of the sensible heat required to heat the juice up to the boiling point and the finite temperature drops. The flow diagram is very complicated to achieve a high degree of economy and heat integration (Figure 5).

The other 50% of the orange is by-product. The peel is dried and used as an animal feed and the waste waters are concentrated and create a material called citrus molasses. The oil is recovered from the peel and is sold as a flavor ingredient. There is oil recovered from waste water, which is also concentrated, purified, and sold. There are other products that are smaller in volume, such as citrus flour, which is dried rag, albedo, and pulp fragments. It has a very bland flavor and is a good thickening agent in foods. Some of the small fractions of other chemicals in citrus are unique. Several of them, one from grapefruit and one from orange, can be turned into unusual artificial sweeteners. It is not as good a sweetener as aspartame, but it is an interesting chemical, called Neohesperidendihydrochalcone (3).

REFERENCES

1. Agricultural Research Service. *Chemistry and Technology of Citrus, Citrus Products, and By Products.* Handbook 78, p. 16, U.S. Government Printing Office, Washington (1962).
2. Bartholomai, A., ed., *Food Factories*, VCH, New York (1987).
3. Clark, J. P., Dihydrochalcone sweeteners, *American Soft Drink Journal*, 30 (1971).
4. Gilson, P., *Industrial Market and Energy Management Guide, Food and Kindred Products*, American Consulting Engineers Council, Washington (1985).
5. U.S. Department of Agriculture, *Agricultural Statistics*, U.S. Government Printing Office, Washington (1989).
6. National Food Processors Association, Hyperfiltration as an Energy Conservation Technique in the Food Industry, DOE/ID/12227-T1 (1984).
7. Renshaw, T. A., S. F. Sapakie, and M. C. Hanson, Economics of Concentration Processes in the Food Industry, presented at AICHE National Meeting, New Orleans (1981).
8. Strobel, R. G., Orange juice concentrate, U.S. Patent 4,374,865, February 22, 1983.

9
Food Product Development

Some of the most interesting challenges to food process engineers occur in connection with the development of new products. Product development is almost always a team effort involving professionals in disciplines which may be unfamiliar to an engineer. A new and strange vocabulary is used, decisions appear to be made irrationally, and time pressures are intense. Although there is no substitute for direct experience, this chapter should help reduce some of the apparent mystery of this critical area.

I. WHY NEW PRODUCTS?

A number of forces motivate new product development. These include the following:

Replacement or rejuvenation of existing products that have declining sales
Responses to identified societal trends
Responses to competitors
Exploitation of new technology
Expansion of sales by extension of existing lines
Repositioning of existing products
Filling of identified needs

Most consumer products, indeed most products of any kind, seem to have a finite life span, almost like a living organism. Their vigor, as reflected in unit sales, much like our own vigor, initially is tentative, then enjoys rapid growth, usually followed by a period of relative stability and, eventually, decline. Some exceptional food products enjoy lifetimes of which humans would be proud—75 years for Twinkies (a filled sponge cake snack sold by Continental Baking Company), for instance—but most survive for much briefer periods of time. For a corporation to thrive, it must constantly replace the sales lost by decline or failure of its existing products as well as generate new sales volume. Since many new products fail to achieve satisfactory sales, it is necessary to develop more new products than would otherwise appear sufficient just to maintain overall corporate sales, let alone achieve any growth.

Societal trends can create opportunities for successful new products. For example, more women work outside the home in the 1990s than in the 1950s; family size is smaller than it used to be; people are living longer; there is a new awareness of nutrition and the impact of food on health. These trends, and others, have resulted in such developments as frozen prepared meals, single-portion products (as compared with family size packages), reduced sodium, calorie and fat products, high fiber products, and many others. Ordinarily, it is the responsibility of the marketing strategists in a company to identify such trends and to relate the firm's new product effort to them, but it is important that the engineers and scientists who must execute the strategy understand the motivating forces.

Every firm, indeed, every product, has some type of competition for the consumer's attention and purchase dollar. When a competitor introduces a new product or new feature, it is necessary to counter with some response. This may be a different product, an imitation, or some new feature. In any event, it requires a product development effort.

New technology may make a new product possible or create a new opportunity. A good example from the mid-1980s is the microwave oven, a household appliance which responds well to current lifestyles by permitting rapid food preparation. Astute food companies observed that conventional prepared foods did not perform well in microwave ovens because they often used aluminum packaging and were not formulated to respond properly to microwave energy. On the other hand, until microwave ovens were found in a large fraction of households, there was too small a market to justify specially designed products. For some companies, the solution to the dilemma was products with dual oven capability, that is, products in plastic packaging which could withstand conventional oven heating. For others, it was products specifically designed for microwave oven use, for example, cake mixes, popcorn, and pizzas. An important aspect of these special designs was new packaging materials which function as conductive surfaces by absorbing microwave energy. This permits crisping of pizza crusts, for instance. Other examples of technology-driven new products abound: soluble coffee and tea, aseptically packaged fruit juices, extruded ready to eat breakfast cereals, quick cooking rice and many others.

Building on existing products, by addition of flavors or features, is known as line extension and is probably the most common approach to new product development. In marketing, it is recognized that an existing, accepted brand name is immensely valuable because the brand is a symbol, carrying with it complex and important significance to consumers. The meaning of a brand is built up over many years by advertising and by the success of products sheltered under its umbrella. There is some debate over whether a brand's significance is diluted when products are added or whether the shelter can easily be extended. Examples can be found to support either case. In any event, building on an existing brand and product line is almost always easier than creating a new brand from scratch.

The degree of line extension can range from simply adding new flavors to a line of soups, ice cream, or cereals to completely new products or new forms of older products with the same name. For example, General Foods has taken both approaches in its line of desserts: new flavors of Jello-powdered gelatin and pudding desserts are constantly introduced; at the same time, totally new ready-to-eat versions have been developed using the same familiar names. Jello Pudding Pops are an example of a line extension well beyond the product definition and are really a new concept (frozen novelties) with a well-known old name.

Repositioning of existing products may be achieved solely by new advertising, marketing, or packaging, but it can also involve some science and engineering. For example, cookie manufacturers have successfully changed the perception of their product from a dessert to a snack, in part by offering the product in smaller package sizes. They have also expanded their markets by creating "soft" cookies in imitation of homemade cookies.

Ideas for repositioning or other approaches to new products may come from consumers. For instance, after dry soup manufacturers learned that people often used their dry onion soup mix to make dips, they developed other formulas specifically for use as ingredients. When tomato sauce manufacturers found that consumers added vegetables, such as peppers, onions, and mushrooms to make pasta sauces, they began to offer new sauce products with vegetables, meat, and special seasonings. As previously mentioned, ready-to-eat versions of foods previously sold as mixes, are becoming more common in response to a perceived need for convenience.

Finally, as new needs are identified, foods, will be developed to address those needs. Achieving restaurant quality food in the home without having to cook is a recently perceived need. One response is refrigerated prepared meals, which have shorter shelf lives than frozen or canned foods, but also have significantly higher quality. Another response may be home delivery of centrally prepared foods, perhaps by visionary supermarkets, which already are offering take-out foods in great variety. A prototype for such an approach exists in the "Meals on Wheels" system for feeding home-bound elderly people and, perhaps, in franchised pizza shops with rapid delivery systems.

Perceived health and nutrition needs create demand for new products and new technologies, such as increased fiber content, non-nutritive sweeteners, calcium fortification, reduced nitrites, and reduced sodium. Products with these features create challenges in formulation and processing. For instance, nitrite has been a reliable food preservative for bacon, corned beef, and lunch meats, but it also can promote the formation of carcinogenic nitrosamines in some products under certain conditions. One approach is to replace nitrite with other preservatives, including salt, but that raises the sodium content and really produces a different product. Another approach is to reduce the nitrite content and rely more heavily on refrigeration for preservation. In the latter case, products may be inoculated with harmless microorganisms which will grow only if the product is somehow abused, as by holding at high temperatures. Such growth will produce lactic acid, which reduces the risk of pathogenic organisms and also alerts the user to the abuse.

Taken together, the motivations mentioned here and others lead to certain generalizations about current trends in new food product development. Convenience, quality, and healthy are probably the three most common descriptors. Convenience is manifested in such characteristics as: microwavable, shelf stable, ready to eat, refrigerated, and single serving. Quality is reflected in less harsh preservation techniques, up-scale or gourmet formulations, ethnic influences, and more expensive ingredients. Finally, the health influence is seen in reduced calories, reduced sodium, increased fiber, whole grains, fewer additives, "natural" ingredients, and even reduced serving size. The food process engineer, as a key member of the multidiscipline team responsible for new product development, needs to understand these confusing and often conflicting influences.

II. HOW DOES PRODUCT DEVELOPMENT OCCUR?

Often, the concept for a new product is generated by specialists in marketing, who may or may not have some technical education. They observe the trends and influences previously discussed, consult with consumers and outside experts, and respond to the overall corporate strategy and business plan. They present their concepts to the corporate or division research department, or to an outside laboratory, with the request that prototypes be prepared, usually, from the researcher's point of view, in an unreasonably short period of time. On other occasions, the concept may originate in technical research and be presented to marketing for a reaction. Product development is an iterative process alternately driven by marketing and research. In any event, the first encounter between research and marketing is only one of many occasions for a clash in culture between the objective scientific method and the subjective marketing methodology.

The usually intense time pressures arise from circumstances over which neither research nor marketing have control, namely, the relatively limited windows of opportunity for advertising of new products on network television, the most commonly used medium.

Unfortunately, process engineers are rarely involved in the first stages of product development, where the emphasis is on fabrication of prototype samples which approximate the image in the mind of the inventor. Sometimes there may be a competitive product which is to be matched or improved upon, but often there is only an idea. The engineer's absence at this stage is unfortunate because directions may be taken which can greatly complicate the manufacturing task later, whereas an engineer on the development team can anticipate the consequences of fabrication decisions and either offer alternatives or begin solving problems earlier.

It also is helpful for the process engineer to understand the critical characteristics of the new product and how they are evaluated, because it is essential that these be preserved through the scale-up exercise. Typically, food products are assessed subjectively by people who may or may not be specially trained to detect certain characteristics. Where possible, objective evaluations are also made, but it is hard to measure objectively such characteristics as flavor, aroma, and mouthfeel. Thus, panels of people become the measuring instrument of the product development team. Sensory evaluation using panels of people must be carefully arranged and the results subjected to statistical analysis.

The number of possible variables in a reasonably complex food product is so large that only a few can be explored systematically. One of the talents of a good product development scientist is knowing which variables deserve the most attention. Realistically, some of the earliest and most critical evaluations are the highly subjective responses to prototypes of the marketing team members and management executives. Although such people may be untrained as sensory evaluators, they usually have good instincts about consumer preferences. However, their personal tastes may or may not be representative of the market; the development team must test repeatedly against larger and larger samples of people to get useful direction.

Thus, with or without a process engineer, the initial stages of product development proceed by trial and error, guided by sensory evaluation panels and subjective comments from marketing people and executives, until one or more acceptable prototypes have been prepared. Usually some key parameters remain to be established, such as serving size, alternative ingredients, packaging and, perhaps, processing conditions. The prototypes may not have been prepared on typical production equipment, because of the low volumes required and the usual unavailability of representative devices.

At this point, it is common for process engineers first to become involved, as the need arises to have larger quantities of test product. In many companies, pilot plants exist to manufacture small quantities of new products under conditions approximating full scale plants; in other firms, actual plants must be used to manufacture test quantities. This latter procedure is typically very expensive because of the relative inefficiency of making small quantities on large equipment; however, it can help expedite subsequent manufacture by surfacing production problems early.

Throughout the development process, the process engineer's focus is on obtaining quantitative data that will permit design or modification of appropriate equipment. In contrast, the food scientist responsible for the product is most concerned with the taste, color, texture and other characteristics of the product, with little regard for exactly how these will be maintained on a large scale. The engineer's experiments may well, and probably should, often produce inedible or unrecognizable products, because the engineer is exploring the limits of process conditions. The product developer would usually view an experiment whose result was inedible as largely wasted time. The culture gap between individuals trained as food scientists and those educated as engineers can be nearly as great as that between the technically trained and the marketing and business people. An understanding of this potential for poor communication can help greatly to prevent it.

As the scale of testing increases beyond the walls of the laboratory, the costs and risks also increase dramatically, but certain aspects of the marketing program can only be tested in the real world. There is an escalating sequence of consumer tests that yields increasingly useful information at a rapidly increasing cost. Early in product development, it is common to use focus groups, which are small, carefully controlled and led groups of consumers who are invited to talk about a product category of concept. From these discussions, researchers and marketing people obtain ideas which may lead to new products and product improvements. Leaders are specially trained professionals, often outside consultants. The discussions are often held in rooms equipped with one-way mirrors so the sessions may be observed by interested corporate personnel, including laboratory researchers and engineers.

Once some quantities of prototypes are prepared, it is common to try samples on consumers picked more or less at random in some convenient location, such as a shopping mall. This is called a central location test and is usually performed by an outside agency. The manufacturer may or may not be identified, since competitors could easily learn of such activities and gain an insight to new product plans. The engineer's involvement at this phase is probably focused on preparation of the samples for testing, but the engineer should also be alert to the impact of consumer reaction on changes to the product and process.

If central location testing (CLT) is encouraging, or after changes are made in response to such tests, the next usual step is home use testing (HUT). Here samples prepared and packaged in near commercial fashion, but often with no corporate identification, are provided to consumers for use in their homes. Information is gathered by questionaire and interview and statistically analyzed. One of the most critical questions is whether the user would buy the product at the anticipated price. The answers to such questions can be misleading, but they are considered critical in deciding whether to proceed. Up to this point, the investment in new product development is generally relatively low; the next step, test marketing, is substantially more expensive.

Whereas samples for CLTs and HUTs probably can be prepared in a laboratory or pilot plant, the quantities required even for a relatively small test market probably preclude this. The food process engineer on the team must either find a way to produce product in an existing facility, build a new facility for the purpose (an expensive and time-consuming proposition) or have the product prepared elsewhere. This last option, often overlooked, is called co-packing and relies on a large network of independent and largely little known firms which manufacture other companies' products for a fee. Although experienced co-packers are usually quite flexible and efficient, a client firm and its engineers must work very closely with the co-packer to maintain quality control and protect proprietary information. Concerns about these issues are significant when manufacturing occurs outside of company facilities, but such concerns can be managed.

In parallel with development of the physical product, a marketing and advertising program is also developed. The first goal of most advertising is to motivate trial of a product; the best product in the world will not succeed if no one tries it. Repeat purchases, without which no product survives, depend on the product and

how well it addresses the consumer's needs. Thus most test market
trials are designed to measure both the effectiveness of the advertising and the reception of the product.

Television is the most popular, though not the exclusive, medium
for food product advertising. Television ads, especially for new
products, are most effective at certain times of the year, such as
the beginning of fall, when new shows typically debut and children
return to school, and in spring, before reruns begin and while
people are planning ahead to vacations. Major holidays are distrac-
tions, and certain products have strong seasonal appeal. The result of these influences is sharply to limit the periods of the year
in which firms like to introduce new products. In turn, these limitations create the severe time pressures previously mentioned. Working backwards from a fall introduction, allowing time to order packaging material, prepare advertising, and ship adequate quantities to
stores, a new product probably must be completely defined by winter of the preceeding year. In the typical case, when decisions on
strategy may be delayed until late in the calendar year, which often
is the fiscal year, there may be only a few months to complete the
entire development process.

In addition to advertising and the product itself, there are other
parameters to be tested: sales terms to stores (suggested selling
price, whether it is preprinted on the package, arrangement in the
store, markup, promotional discounts, advertising allowances, stocking fees, minimal order quantity, access to sales data and other possibilities), consumer promotions (discount coupons, door-to-door samples, in-store samples, trial sizes, and others), and distribution approach (brokers, company sales force, store door delivery, stocking
distributors). Some possibilities are dictated by the product (whether
it is perishable or not, for instance), whereas others are testable,
using trial, repeat purchase and financial returns as measures of
success. Test market locations are typically dictated by the overlap of television station coverage with distribution limits of chain
stores. Desirable test markets are relatively isolated, medium-
size cities in which television advertising is relatively efficient and
quantities of product required are reasonable. Well-known examples include Columbus, Phoenix, Atlanta, Syracuse, and Kansas
City. Large cities are too difficult to supply and inefficient and
expensive for advertising.

III. SENSORY EVALUATION

It is helpful to understand some of the inherent limitations of sensory evaluation, because it is so critical in food product development and yet it is often misused. The human response to food

involves nearly all the senses: taste, of course, is most important, but so also are smell, sight, feel, and even sound. It is difficult to separate the responses. For instance, most people would find unpleasant plain milk dyed green or blue, even though the dye contributes no taste. To counter such influences, properly designed sensory evaluation facilities have lighting systems which can be controlled to mask colors, if desired, or to simulate various lighting conditions, such as daylight or store-like artificial illumination, so that the visual response will be uniform among testers and not a source of confounding.

In sensory evaluation, the usual goal is to identify preferences among alternative formulations or treatments, using a small panel to represent the consuming population. A typical approach is to rank samples on a hedonic scale (1 to 9, for example) where five is a neutral reaction, nine is usually "like very much" and one is "dislike very much." Such a ranking alone provides little information about the reasons for liking or disliking. Often a panel is asked to make additional comments, which can provide useful guidance, but is difficult to quantify.

Different information is generated when panels are asked to distinguish among samples, usually in a triangle test, where two samples are identical and one is different. One function of such tests is to calibrate a panel or its individual members as to their sensory acuity, by measuring the concentrations of some substance (such as a flavor) at which most people fail to distinguish a difference from distilled water. When measured with a large number of normal people, such threshold concentrations are reliable characteristics of a given compound. (Some flavor compounds are remarkably potent, detectable in parts per billion.)

One result of such testing is the discovery that some people are "blind" to certain tastes and odors. Such a reduction in sensory acuity may be genetic or it may be the result of disease or medication. Smoking reduces, but rarely eliminates sensory acuity. The discovery of true genetic "blindness" for certain flavors and aromas, rare though it is, suggests the existence of "primary" tastes and aromas, analogous to the primary colors of sight (for each of which there are known color blindnesses). The primary tastes are considered to be salt, sweet, acid, and bitter. There are many more candidate primary aromas, perhaps as many as 20.

In addition to the fundamental complexities of taste and smell, there are additional complications due to age, injury, disease, medication, and cultural influences. General acuity is thought to decline with age, but the effect is not necessarily uniform across the spectrum of tastes and smells nor among all people. Blows to the head can sever olfactory nerves and affect the sense of smell. Anyone who has suffered a common cold knows the effect of one disease on reducing sensory responses. Other diseases and conditions, such as some mineral deficiencies, affect taste and smell. Finally, some flavor preferences are acquired and can vary among cultures; common examples include hot pepper, garlic, cumin, and ripe cheese.

Unless a new product is directed at a specific audience, such as the elderly, it is probably best to satisfy the indicated preferences of a panel representative of the average American public, or, perhaps, of a panel representing some region. There are regional differences in taste preference, and increasingly, large firms are attempting to satisfy such regional preferences. However, identifying such preferences with a panel from another region does not work! Triangle tests for acuity can help calibrate a panel as "normal"; preference tests for known regional stereotypes can calibrate for regional accuracy.

Generalizations that have been accurate in the past, for processed foods directed to the national market, indicate that Americans like sweet, salty, and relatively bland (unspiced) foods. However, some of the trends discussed earlier seem to support the impression that tastes may be changing. The increasing popularity of such ethnic categories as Italian, Mexican, and Oriental foods suggests an increasing tolerance for more exotic flavors, aromas, and textures. One impact on the product development team will be the need to explore more widely in the already complex domain of sensory response.

The significance of all of this to the process engineer is to realize that his or her experiments cannot focus exclusively on temperatures, flows, and pressures, which are relatively easy to measure accurately, but also must consider the messy and hard to measure variables of taste, aroma, and texture, to which humans have emotional and unreliable responses.

10
Scale-Up of Food Processes

I. WHAT IS SCALE-UP?

Scale-up is a poorly defined and even more poorly understood concept. Little has been written about the general solution to scale-up problems. This is particularly true when the chemistry and physics of the problem are difficult to define, as is often the case with food processing.

The objectives of this chapter are to define the dimensions of the scale-up problem and to illustrate possible approaches to achieving practical solutions to real problems.

A. Definition

The dictionary states that scale-up is an increase in size at fixed ratio. While this definition has something to do with the problems facing the design engineer, it clearly does not adequately define the problem. A better working definition might be given as follows:

Scale-up is the task of producing an identical, if possible, process result at a larger production rate than previously accomplished.

Although not stated, this definition assumes that the increased production capacity will be accomplished with equipment that is physically larger than that previously used. So, in the context of this book's material, we will not specifically be speaking about simply increasing the speed at which fixed equipment operates, although the two problems are admittedly related. If sufficient understanding

exists to successfully design larger equipment, the engineer probably will understand the consequences of increasing speed.

B. Concept Development

Scale-up is not choosing larger equipment from catalogs. Sad to
say, this is what often happens in the food processing industry.
One should beware of suppliers "scale-up" techniques. Experience
teaches that suppliers generally have less understanding of a product than the manufacturer.

   Although not generally recognized, process scale-up is an everyday occurrence, which has been successfully executed for hundreds
of years. Consider the statement found in almost any cookbook:
"Cook the turkey for 30 minutes per pound." In a sense, this is
a resolved scale-up problem. Someone has determined that to increase production rates, for example, more meals per batch, the
time to obtain the same extent of cooking increases with the size
of the object being cooked. This has been quantified by a "scale-
up rule," "cook 30 minutes per pound."

   The cook who established this simple rule did so by observing
the preparation of many individual objects. Because of the high
cost of processing equipment and experimentation, this is generally
not a viable approach for process development. Therefore, the object of any scale-up experimentation should be to create a scale-up
rule with a minimum of experimentation. Fortunately, a knowledge
of engineering principles allows this to be accomplished.

   Consider what the cook has done:

A desired process result has been defined. In this case it is the
   extent of cooking or internal temperature of the turkey.
Some process description has been defined. In this case it is the
   cooking time per pound. The process description is the design
   criterion that, presumably, yields the same extent of cooking or
   internal temperature of the turkey, independent of the roast's
   size.

   This leads to a key definition:

The *primary scale-up criterion* is that process parameter, or set of
   process parameters, that brings about the desired process result independent of the ultimate scale (size) of the process.

   The task then is to clearly understand this statement's meaning and
application to problems of any degree of complexity.

   The example can be generalized for any sort of scale-up problem. The process could be the cooking of a piece of meat, the mixing

of ingredients in a tank, or the puffing of a product with a cooking extruder. For any of these processes, the development engineer can run a series of experiments in which the effect of varying some condition(s) is used to create different process results (product attributes). The results might look something like those illustrated in Figure 1.

One conclusion might be that scale-up is as straightforward as choosing the value of the process parameter that gives the same or desired result. Following this incorrect line of reasoning, process scale-up might be considered as simple as choosing the same value of the process parameter for the larger process. A commonly observed example of this fallacy is mixing a product for 10 minutes in the pilot plant at some speed and then incorrectly using the same mixing time and speed for the full-scale process.

This mistake will not occur if the definition of the primary scale-up criterion is followed. Implicit in the statement "independent of scale" is the requirement that proper scale-up requires experimentation is more than one scale.

If the example is taken one step further, and recognizing that more than one scale of experimentation is required, additional experiments should be run on a larger, though not full, scale. The experimental results might now appear like those illustrated in Figure 2.

Remembering that the goal is to design a process of still larger size, Figure 2 does not yield a viable definition for the primary scale-up criterion. The key words in the definition are "independent of scale." The experimental result shown in Figure 3 is what is needed.

Figure 3 shows a process parameter yielding a result that is independent of scale. In order to design a third, larger, scale of the process the engineer can be reasonably confident that designing the new scale while keeping the process parameter constant will yield a product having the desired properties. This is not certain, but the confidence level can be greatly increased by judicious experimental design and evaluation.

What can the process result be? In fact anything we choose it to be. Some examples are presented in Table 1. There is, of course, much to be desired in moving away from the very subjective, nonquantitative types of process results listed at the top of the table, but there is no fundamental reason why this type of process result cannot be used. Engineers in the food industry are often forced to use these types of qualitative process responses. This is a direct consequence of having to deal with poorly understood chemistry and physics.

The process parameters may be any rational, quantitative measure of the process state. All that is required is that the chosen

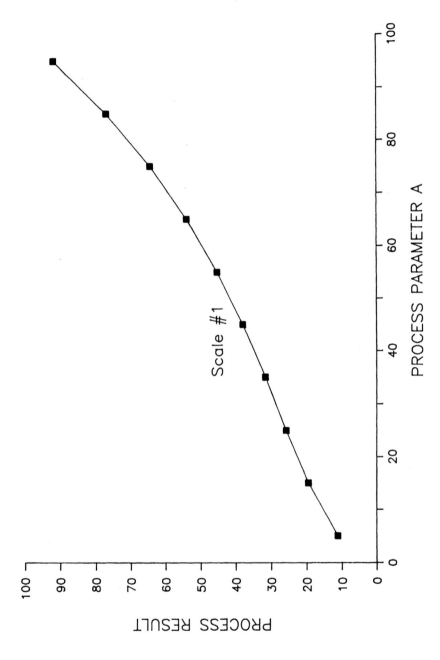

FIGURE 1 Experimentation on only one scale.

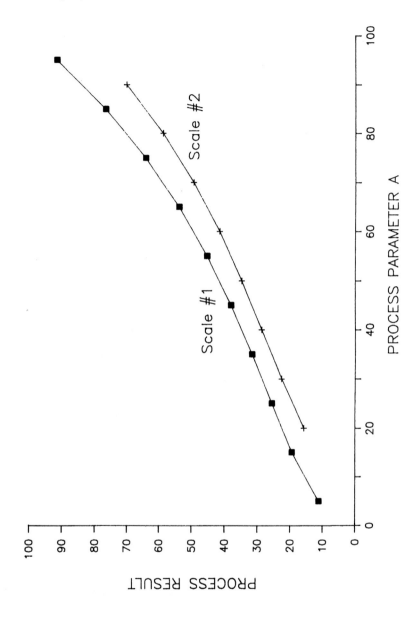

FIGURE 2 Experimentation on two scales.

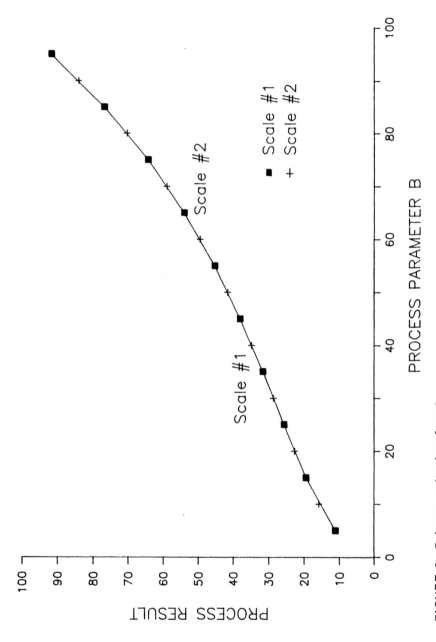

FIGURE 3   Primary criterion found.

TABLE 1   Some Examples of Process Results

Subjective	Objective
Flavor	Reaction rate
Texture	Heat transfer rate
Stickiness	Mass transfer rate
Overall acceptability	Selectivity
	Dimensionless numbers
Fundamental	
Temperature	
Pressure	
Concentration	
Flow rate	
Viscosity	

parameter be measurable or somehow quantifiable.   Some examples of process parameters are presented in Table 2.

Engineers frequently use the general concept that has been described but often are not cognizant of the general need for experimentation at more than one scale.   The data shown in Figure 4 illustrate this point.

TABLE 2   Some Examples of Process Parameters

Fundamental
  Temperature
  Pressure
  Concentration
  Flow rates
  Velocity

Calculated or measured
  Tip speed
  Power
  Torque
  Power/unit volume
  Energy/unit volume
  Shear rate
  Total strain
  Circulation rate
  Superficial velocity
  Circulation/velocity head
  Dimensionless numbers

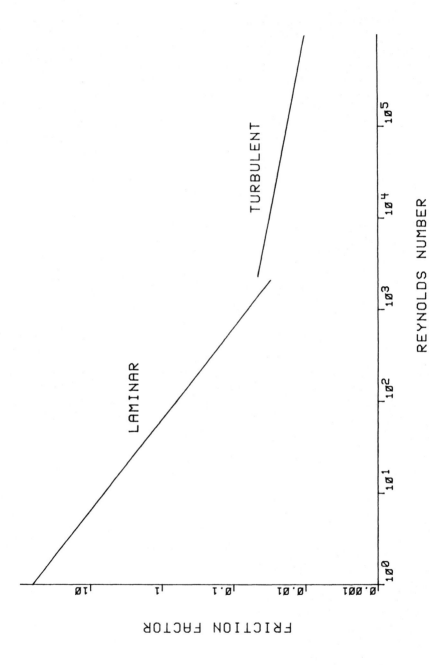

FIGURE 4   A familiar scale-up rule.

In this case the process result is the dimensionless friction factor. Physically, the friction factor represents that fraction of kinetic energy which is lost to friction. The process parameter is the Reynolds number. This relationship is accepted to be true for all sizes (scales) of pipes. In fact, these curves have a theoretical basis for the laminar flow regime and, until long after their development, were empirical in the turbulent regime. Originally, the correlation was developed by showing that the data for many sizes of pipe fit on these universal curves. Other examples, accepted as being true for all scales, are heat transfer correlations of the form

$$Nu = f(N_{Re}, Pr) \tag{1}$$

where Nu is a Nusselt number; N_{Re}, a Reynolds number; Pr, a Prandtl number.

In this case the process result, the Nusselt number, is a way of describing the heat flux. The scale-up is defined as a function of the Prandtl and Reynolds numbers. Most correlations of this type are established empirically.

C. The Turkey Problem: An Example

1. Discussion

To illustrate what experiments and data evaluation are required, consider the cooking of a turkey. Assuming that the properties of the turkey are isotropic and that the turkey is roughly spherical, the problem can be solved. Assuming that the initial temperature of the turkey and the oven temperature are to be held constant and that the external heat transfer resistance is negligible, the relationship between temperature, time, and turkey size can be reduced to Eq. 2 from the fundamental differential equations for unsteady state heat transfer (4,11,12,18,20,25,39,49):

$$T = f(t/D^2) \tag{2}$$

In terms of the mass of the turkey,

$$m \propto D^3 \tag{3}$$

making

$$T = f(t/m^{2/3}) \tag{4}$$

where

T = the center temperature of the turkey
t = the cooking time
D = the characteristic diameter of the turkey
m = mass of the turkey

If the final center temperature is specified, Eq. 4 can be simplified into a simple scale-up rule.

$$(t/m^{2/3}) = \text{constant} \tag{5}$$

The process parameter sought is the ratio of cooking time to the
two-thirds power of turkey mass. The validity of the scale-up rule
can be verified by converting the cooking times printed on turkey
labels to the parameter defined by Eq. 5.

This is a case where experimentation at more than one scale is
not required because the underlying physics is well understood.
The problem can be solved with a minimal understanding of the
physics of the problem. This is the case for most food processes.

To illustrate this point, assume that the only knowledge available is the fact that the quantity of heat required to cook the turkey is in some manner proportional to the mass of the turkey. Now
consider several hypotheses about how a turkey cooks or how heat
is transferred into the turkey.

Hypothesis 1: Radiation energy from the oven strikes the turkey
  so that all the energy is absorbed. This type of hypothesis
  might arise from observing the "gross" behavior of a microwave
  oven.
Hypothesis 2: The quantity of heat absorbed is proportional to the
  turkey's surface area.
Hypothesis 3: The movement of heat is analogous to the flow of
  electricity.

Other hypotheses that might describe more or less realistic models for heat transfer are possible. The three hypotheses given
above are sufficient for illustrative purposes. The next task is to
convert the hypotheses into mathematical expressions suggesting
process parameter candidates that can be checked against experimental data.

Hypothesis 1 states that all energy radiating from the oven walls,
coils, or burners is absorbed by the turkey. If this is correct,
then the time to cook the turkey is the heat required divided by
the heat input, which is the power output of the oven.

$$t \propto m/P \tag{6}$$

or

$$(t/m) = \text{constant} \tag{7}$$

Hypothesis 2 states that the rate of heat absorption is proportional to the area available for absorbing heat. The time to cook the turkey is then proportional to the mass of the turkey divided by its surface area. The surface area is proportional to the square of turkey diameter. This parameter may be written as:

$$t \propto m/D^2 \tag{8}$$

or

$$(t/m^{1/3}) = \text{constant} \tag{9}$$

Hypothesis 3 relates the behavior of heat movement to the behavior of electrical flow, a frequently used analogy. In this case, the flow of heat is described as the ratio of driving force to a resistance. Electrical resistance is defined as the product of the resistivity of the material times the length of the material divided by the area through which the electricity flows (21,55). Using the same analogy for heat "resistance,"

$$r \propto D/D^2 \tag{10}$$

where r = the "resistance," or

$$r \propto m^{-1/3} \tag{11}$$

The "driving force" is simply some temperature difference between the turkey and the oven. For this problem the average driving force is approximately

$$\text{Driving force} = [(T_{oven} - T_i) + (T_{oven} - T_f)]/2 \tag{12}$$

Since the final temperature has been specified the driving force is a constant.

As before, the total heat transfer requirement is proportional to the mass of the turkey. So the time to transfer this heat is

$$t \propto mr \tag{13}$$

or

$$(t/m^{2/3}) = \text{constant} \qquad\qquad (14)$$

The tasks required to establish the primary scale-up criterion are almost complete. Suppose the ultimate process scale will be a very large turkey, which is unavailable.

Experiments are run to establish the cooking time for two or more small turkeys of different size. Cooking times for each experimental turkey are converted to the hypothesized process parameters. The parameter that has the same value for both experimental turkeys is the primary scale-up criterion. Note that this is a one-dimensional analog of Figure 3.

The following data were copied from the label on a stuffed turkey from a supermarket.

Turkey weight (lb)	Cooking time (hr:min)
8–12	3:30–4:30
12–16	4:30–5:30

Using the middle of the specified ranges as the data, the following data is obtained for the primary criterion candidates.

Turkey weight	Parameter for hypothesis		
	1	2	3
10 lb	0.400	1.86	0.862
14 lb	0.357	2.07	0.861
Percent difference	10.8	11.2	0.1

Hypothesis 3 is the only one that makes the process result independent of process scale. The primary criterion has been established for this process.

$$t/m^{2/3} = 0.86 \qquad\qquad (15)$$

If the data for temperature versus cooking time had been provided, then temperature could have been plotted versus the candidate parameters and the result would look like Figure 3 only for the third, correct, hypothesis.

2. Lesson Learned

The key points of this example are as follows:

1. The correct method for scaling up a process can be determined even without resorting to an in-depth analysis of the underlying principles.
2. If data were presented as an organoleptic (extent of cooking) rating of the turkey versus cooking time, the same procedure could be followed to find the correct relationship between cooking time and weight that would yield a constant organoleptic result.
3. All that is required is well thought out experimentation at more than one scale.

Of course, this particular problem could have been solved through the use of dimensional analysis. The same result would be obtained, and only one experimental scale would be required. Unfortunately, the desired process result for most food processes is a not as readily quantified property such as temperatue, flow rate, or energy. Instead, the normal desired result is organoleptic response, or some other property, such as cake volume or texture, which is not readily related to the underlying physics of the problem. In fact it can be argued (6,27,30,58) that semiquantitative measurements of this type violate one of the basic premises of dimensional analysis: the requirement that the variables have the property of absolute significance of relative magnitude. In later chapters there will be a detailed discussion of the application of dimensional analysis.

Before proceeding further with "real" examples, there are other requirements of scale-up which do not appear in the turkey cooking problem. In a real process one must specify additional aspects of process design which change as the process is scaled up. These are often mechanical items, which are necessary to fulfill the primary scale-up criterion. For example, if dough is being mixed, the dough will heat up via mechanical energy conversion. If the dough temperature is critical to the dough quality, the relationship between mechanical energy input and external heat transfer and process scale must be known. This leads to the following important definition, for secondary scale-up criteria.

D. Secondary Scale-Up Criteria

Secondary scale-up criteria are those mechanical and physical changes with scale that must be understood in order to satisfy the primary scale-up criterion. Table 3 illustrates some obvious, and not so obvious, secondary criteria which must be developed to properly scale

TABLE 3   Some Examples of Secondary Criteria

Effect of scale on heat transfer

Effect of scale on horsepower or torque

ASME code

Shaft design

Integral sizes

Physical deflection

a process. It will be shown that fulfillment of the secondary criteria is often impossible and thus the ultimate scale of the process
which can be designed without compromising the desired process result is limited.

E.   Scale-Up Process Summary

To correctly scale up a process, one must:

Define the desired process result.
Define the primary scale-up criterion as that parameter or set of pa-
    rameters which makes the desired process results independent of
    scale. This generally requires experimentation at more than one
    scale.
Define the secondary criteria for process scale-up.

It has been stated that the primary scale-up criterion can only
be determined by experimentation on more than one scale. This is
essentially true, with exceptions occurring when the physics of the
process is very well understood. This is generally not the case in
food processing operations. There are exceptions, some of which
are listed in Table 4.

II.   GEOMETRIC SIMILARITY: TWO-PHASE
       (GAS-LIQUID) MIXING

The problem discussed here illustrates a more complex and useful
problem and demonstrates most, if not all, of the scale-up principles mentioned thus far. In addition, several new principles are
illustrated.

TABLE 4   Some Cases Where Scale-Up Experimentation
Is Not Needed

Trivial
Piping and pumps specification

Inherently independent of scale
Homogeneous reactions
Batch drying of particles
Frying time
Parallel processing, e.g., liquid cyclones

Scale-up is well defined
Many heat transfer problems
Filtration and centrifugation
Distillation
Suppliers have done work[a]

---

[a]Extreme caution should be used in this case. Many
suppliers do not really understand scale-up. Detail
discussions should be undertaken to convince the purchaser that the equipment has really been scaled-up.

Consider a fermentation reaction occurring in an agitated tank.
To simplify the problem, assume the following:

1. The rate of reaction is controlled solely by the transfer of air
   from the gas to the water phase.
2. The system is turbulent.
3. The reaction vessel is fully baffled.
4. The system is isothermal.
5. The system is Newtonian.
6. Air is always provided in excess of reaction requirement.
7. Geometric similarity is to be maintained.

The last item, geometric similarity, is extremely important and requires more elaboration.
     The literature (30,58) often defines geometrical similarity as fol-
lows:

Two bodies are geometrically similar when to every point in the one
     body there exists a corresponding point in the other.

This is a complicated way of saying that the ratio of any two
lengths in the first is the same ratio in the second. This is seen

in the distortion-free enlargement of a photograph. This definition
may be too restrictive. For example, the second body may be made
of a multiplicity of cells which are each geometrically similar to the
original model. An example of this would be sections of a packed
column or multiple distillation plates. In addition, when the physics are understood, the definition of geometric similarity may be re-
laxed somewhat so that similarity is only true in cross section (two
dimensions). As an example, in studying rolling or calendering of
a material, only the cross-sectional view must be geometrically similar on scale-up, if the initial model roll is wide enough for edge effects to be negligible. In the literature (30) this is sometimes called
maintaining the section ratio.

For the gas-liquid mixing problem at hand, the restraint of geometric similarity is illustrated in Figure 5. For even a simple turbine agitation system the mathematical list of geometric restraints is
quite large, as shown in Table 5.

First the desired process result must be defined. Since we are
interested in the rate of reaction, which is controlled by the rate
of transfer of air from the gas to the liquid phase, a suitable process result would be a mass transfer coefficient, defined by

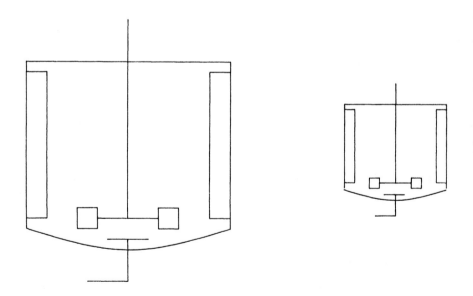

FIGURE 5   Geometrically similar sparged tanks.

TABLE 5 Factors That Must Be Kept Constant in Mixing Systems for Geometric Similarity

Liquid depth/diameter	Number of blades/impeller
Impeller diameter/diameter	Impeller pitch
Baffle width/diameter	Blade width/diameter
Bottom clearance/diameter	Number of baffles
Number of impellers	

$$N_g = k_1 a(c - c^*) \qquad (16)$$

where

N_g = the rate of mass transfer (vol/time-vol)
$k_1 a$ = the product of mass transfer coefficient and interfacial area
c = the concentration of gas in liquid
c^* = the equilibrium concentration of gas in liquid

In order to determine the primary scale-up criterion, first design a series of experiments on two scales. For each tank, there are really only two variables:

1. The volumetric flow of air
2. The agitator speed

Consider a simple set of experiments using these two variables and hypothetical results, as given in Table 6.

Several things are apparent. Speed of the agitator affects the rate of mass transfer but is not the primary scale-up criterion, since the same mass transfer coefficient is not obtained at the same speeds for both scales. In addition, the flow rate of gas has a significant effect on the reaction rate. One would therefore expect that the primary scale-up criterion would be a set containing some description of the condition of agitation and the gas flow.

The problem is to uncover the form of the primary scale-up criterion. This can be done by brute force, statistical methods, or some combination of both. The brute force method is most illustrative and useful for this example.

The task then is to utilize prior knowledge and creativity to develop possible variables which will describe the process. Some examples that might be considered are as follows.

TABLE 6   Data for Gas-Liquid Mixing Problem

Agitator speed (rpm)	Gas flow (cu ft/min)	Mass transfer coefficient (cu ft/cu ft min)
Vessel diameter of 10 inches		
100	100	0.6
	200	1.0
	300	1.3
200	100	4.7
	200	7.3
	300	10.2
300	100	13.8
	200	23.0
	300	29.9
Vessel diameter of 50 inches		
50	2,000	1.5
	5,000	3.0
	8,000	4.2
100	2,000	11.6
	5,000	21.5
	8,000	29.0
150	2,000	35.0
	5,000	67.0
	8,000	112.0

For the agitator:
   Tip speed
   Turnover time
   Circulation rate
   Reynolds' number
   Weber number
   Imparted kinetic energy
   Imparted shear rate and shear stress
   Circulation per unit velocity head
   Power/unit volume
   Torque/unit volume

For the gas flow:
Superficial gas velocity
Gas delivery number (loading factor)
Gas Reynolds' number

The values of these variables at any experimental condition are easily calculated using the information in Table 7. The procedure is graphically illustrated by plotting the measured mass transfer coefficient versus the calculated candidates for the primary scale-up criterion. Figures 6 through 9 illustrate some of the results. In Figure 9 the physics of the process is best described by power per unit volume imparted by the impeller. None of the other parameters give a process result independent of scale. In fact, for most of the candidates, the lines for the two test scales are separated by more than an order of magnitude.

Multiple regression techniques can be used to describe the generalized role of the gas flow rate as illustrated in Figure 10. This conclusion is established in the literature (1–3,17,19,26,50,51,59, 60).

TABLE 7   Some Useful Relationships
for the Gas-Liquid Mixing Problem

Tip speed $\propto ND$

Power $\propto N^3 D^5$

Torque $\propto N^2 D^5$

Velocity head $\propto N^2 D^2$

Impeller Reynolds' number $\propto ND^2$

Weber number $\propto N^2 D^3$

Impeller shear rate $\propto N$

Impeller circulation $\propto ND^3$

Superficial gas velocity $\propto Q/D^2$

Gas delivery number $\propto Q/ND^3$

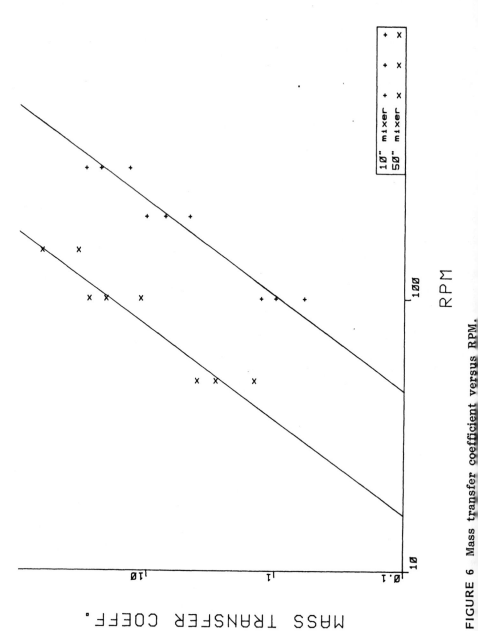

FIGURE 6 Mass transfer coefficient versus RPM.

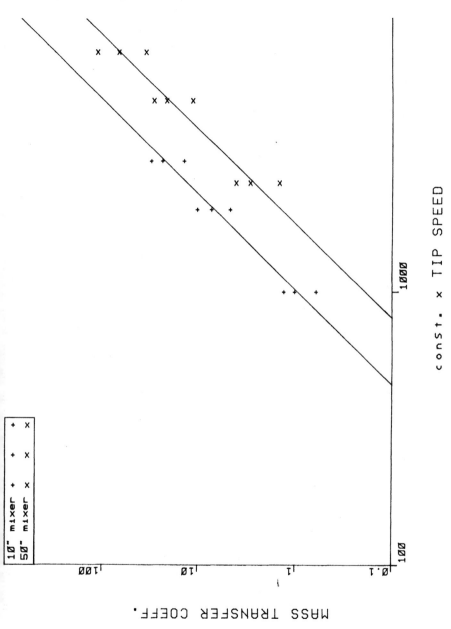

FIGURE 7 Mass transfer coefficient versus tip speed.

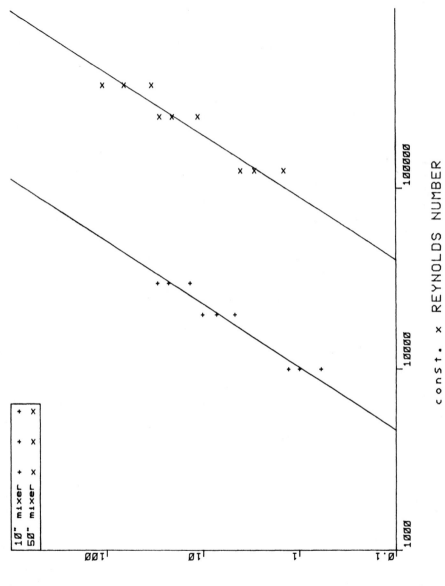

FIGURE 8. Mass transfer coefficient versus Reynolds number.

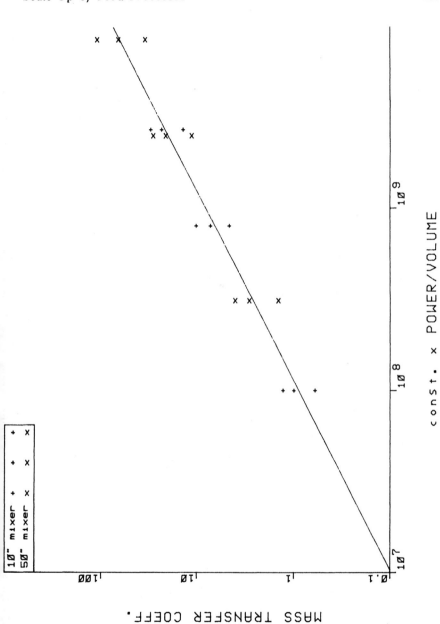

FIGURE 9 Mass transfer coefficient versus power/volume.

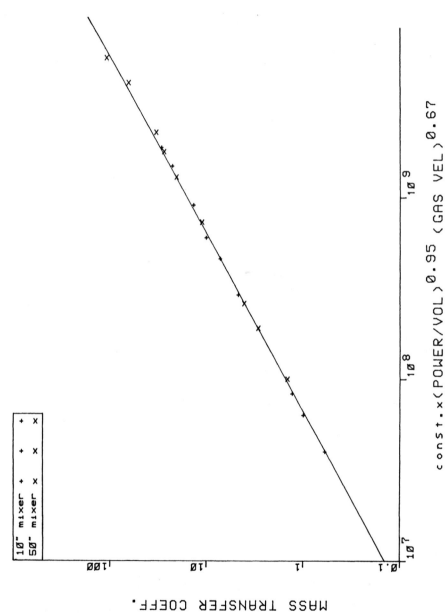

FIGURE 10 Mass transfer coefficient versus a function of power/volume and gas velocity.

A. Potential Consequences of Incorrect Scale-Up

Consider the case of scaling from a 10-inch diameter pilot plant mixer to a 50-inch diameter plant mixer. Suppose that the impeller tip speed was incorrectly chosen as the primary criterion.

Assume also that for the pilot plant mixer a speed of 100 rpm was found to yield the desired reaction rate. For constant tip speed on scale-up to a 50-inch diameter mixer this yields

$$N_{plant} = N_{pilot}D_{pilot}/D_{plant} = 20 \text{ rpm} \qquad (17)$$

The experimentation described earlier showed that the correct scale-up criterion, if we neglect the less important effect of gas velocity, is constant power per unit volume. For the speed determined by assuming that tip speed is maintained on scale-up, the relative power per unit volume for the two scales is

$$(P/V)_{plant}/(P/V)_{pilot} = (N^3D^2)_{plant}/(N^3D^2)_{pilot} = 5 \qquad (18)$$

The data (Fig. 9) indicates that the rate of mass transfer per unit volume is related to power per unit volume by

$$\text{Rate} \propto (P/V)^{0.95} \qquad (19)$$

which indicates that the relative rate of mass transfer between the pilot plant and full scale plant will be

$$\text{Rate}_{pilot}/\text{rate}_{plant} = 4.6 \qquad (20)$$

This does not fulfill the primary objective of scale-up: "process result that is independent of scale."

If a problem of this sort were unexpectedly encountered, the normal approach would be to increase the speed of the agitator. If the error were of the magnitude just illustrated, the required speed increase could probably not be achieved without encountering major mechanical difficulties. These difficulties arise because of some of the secondary design criteria.

B. Importance of Secondary Criteria

Secondary criteria played a direct role in the determination of the primary criterion for the gas-liquid mixing example. The equations presented in Table 7 and in illustrating the consequences of the use of the incorrect scale-up criterion imply a knowledge of the secondary criteria. The most obvious example of this is the relationship between power and speed and agitator diameter.

In this particular example, the relationship can be found in the literature (2,17,51,59), but for more obscure processes this may not be true. For processes where the relationships used for secondary criteria are unknown, the experiments used to determine the primary criterion can also be used to determine the mathematical relationships for secondary criteria.

For the gas-liquid mass transfer example, some of the secondary criteria that must be determined are:

The relationships between scale and speed and power consumption, for both a gassed and ungassed reactor, must be determined.
The gas hold-up of the reactor as a function of scale, agitation conditions, and gas flow rate must be known.
If temperature control of the process is important, relationships between heat transfer and process scale and agitator speed must be determined.
Physical limitations on shaft design must be understood.

The reader could probably add several other design considerations to the preceding list.

Consider the importance of the power/scale relationship. The primary scale-up criterion could have been developed without a quantitative knowledge of the relationship; the power consumption could have been measured at each experimental condition. If this were the case, even though the primary criterion were correctly determined, the information could not be used to design the plant scale reactor. Once the primary criterion establishes that power per unit volume must be kept constant when scaling-up, the mathematical relationship between power, scale, and speed must be used to determine the speed at which the larger reactor must be operated. This leads directly to drive train and shaft design. If the reactor were never operated in an ungassed condition, this would be sufficient. However, this is not a likely occurrence: There is a finite probability of losing gas feed, or the reactor would probably not be gassed until the vessel is charged and heated, or cooled, to the desired operating temperature. The power consumption in the ungassed condition will be significantly higher than in the gassed condition (17,51,59). To prevent mechanical damage to the shaft and drive train, this fact must be considered in a thorough design of the mechanical components.

The realities of the mechanical design are the source of the difficulties that will be encountered when one attempts to correct for a design which used an incorrect primary scale-up criterion, such as illustrated above. If agitator speed must be increased significantly, as in the example, by a factor of 1.67, the drive components and agitator shaft would not have been designed to take the increased load associated with the agitator speed increase.

The importance of the relationship between scale and gas holdup is trivial, but might have, at the very least, an embarrassing consequence. The volume of the reactor contents expands significantly with the introduction of gas!

Heat transfer considerations most often are the secondary criteria which limit the ultimate scale the process can attain. This is easily illustrated.

Assume that a process, such as gas-liquid agitation, has been designed with a primary scale-up criterion that requires that power per unit volume be maintained on scale-up while maintaining the temperature of the mixers' contents. Assume that the reaction has a significant heat of reaction associated with it. For the purposes of illustration, we will assume that the pilot plant conditions which yield the desired process result are 100 rpm on a 10-inch diameter vessel. The desired diameter of the plant reactor will be 50 inches.

In order to maintain constant power per unit volume, the speed of the plant reactor is given by

$$(N^3 D^2)_{plant} = (N^3 D^2)_{pilot}, \quad N_{plant} = 34.2 \text{ rpm} \qquad (21)$$

The rate at which heat must be removed from the reactor is proportional to the volumes of the two reactors. The relative magnitude of required heat removal is

$$\frac{Q_{plant}}{Q_{pilot}} = 125 \qquad (22)$$

The relationship between heat transfer coefficient and reactor scale and agitator speed may be found in the literature (5,50,51,59) or can be determined experimentally. The relationship is

$$h \propto N^{0.8} D^{0.6} \qquad (23)$$

where

h = a heat transfer coefficient
N = the rotational speed
D = the mixer diameter

It follows that the relative magnitude of the heat transfer coefficients is

$$\frac{h_{pilot}}{h_{plant}} = 1.11$$

Assuming that external and wall heat transfer resistances are negligible, the definition of the heat transfer coefficient can be used to determine the relative temperature differences required to maintain both reactors at the same operating temperature.

$$\Delta T_{plant} / \Delta T_{pilot} = (Q/hA)_{plant} / (Q/hA)_{pilot} = 4.5 \qquad (24)$$

This could create considerable difficulties. If the reaction were exothermic and the reaction conditions required that the reaction temperature be near room temperature, the higher required temperature difference for the larger reactor could result in the requirement that the coolant be below the freezing point of the reaction mixture. This in turn could lead to freezing of the mixture on the vessel walls. Conversely, for a heated system the required surface temperature might be high enough to cause thermal decomposition of the reaction mixture.

There are several possible approaches to solving this problem. All have significant economic consequences. Some alternatives are as follows:

Add coils to the reactor to increase available heat transfer area.
Add an external heat exchange loop through which the tank contents are circulated.
Accept a slower reaction rate on the large scale.
Build more reactors of a smaller size than originally desired.

The first two alternatives might present the designer with some new problems:

1. The addition of coils to the reactor interior dramatically alters the geometry of the system. This will result in a change in the local shear patterns within the agitated fluid. In all likelihood this could alter the functional relationship between mass transfer rate and power per unit volume. If this is ignored, the reaction rate in the full-scale reactor will not be that expected. This is a key reason for maintaining geometic similarity of the vessels throughout the initial design phase. *A change in geometry may undermine the proposed technique*[1]. To overcome the problem, a return to the pilot plant is required to evaluate the effect of the geometry

---

[1]This reasoning suggests that not maintaining geometric similarity during the experimentation to determine the primary scale-up criterion could result in an incorrect determination of the primary criterion, or finding no primary criterion. When dimensional analysis is discussed later in the text, a very strong rationale for maintaining geometric similarity will be presented.

on the primary and secondary scale-up criteria. Obviously this adds to cost and time of process development.

2. Although the addition of an external heat exchange loop will probably have little effect on expected mass transfer rates, this may not be a viable technique if the reaction mixture is very viscous. It will certainly add considerable capital cost.

3. For certain processes, for example, extrusion, the additional internal heat transfer surface, or an external heat exchange loop may not be physically realizable.

The importance of heat transfer as a secondary scale-up criterion is considered for extrusion in the chapter on RTE cereals.

III. USEFUL RULES-OF-THUMB ABOUT SCALE-UP EXPERIMENTATION: ADDITIONAL CONSIDERATIONS IN DETERMINING THE PRIMARY SCALE-UP CRITERION

Before presenting experimental examples of the determination of the primary scale-up criterion, it is appropriate to discuss some important aspects of the requirements of experimental design and evaluation of the experimental results:

1. *Keep the experiments as simple as possible.* Design the experimentation in such a way as to minimize the effect of external, additional factors which can have an effect on the process result. For example, temperature could have been added to the study of the gas-liquid reactor. This would have greatly increased the experimentation required, but more importantly, would have complicated the extraction of the primary criterion. In this case, the task is to match a three-dimensional surface for the two scales rather than the two-dimensional line that was required. Once the initial evaluation is complete, it is easy to add some additional runs to examine the effect of temperature on the required value of the primary scale-up criterion. This is essentially what was done by looking at the effect of agitation followed by a consideration of the effect of gas velocity.

2. *Try to eliminate extrapolation.* When a system is poorly understood, as many food systems are, unexpected functionality in the form of nonlinearities might occur. Extrapolation could lead to an incorrect solution, or failure to find any primary criterion. This is avoided by ensuring that there is some degree of overlap in primary scale-up criterion candidates between the two scales being studied. Figures 11 and 12 illustrate the potential consequences of not following this practice. For the nonlinearity illustrated in Figure 11, no primary criterion would be found. In Figure 12, the extrapolation would result in a linear correlation which would be incorrect.

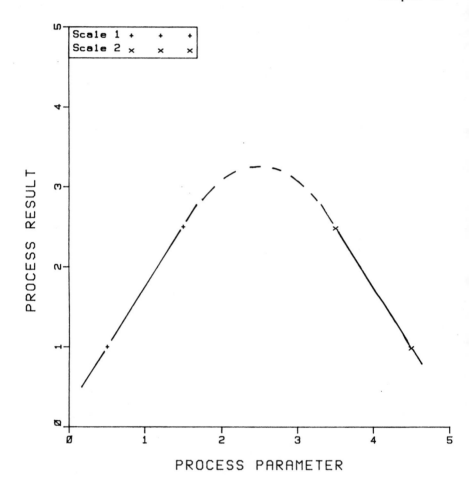

FIGURE 11   A case where a nonlinearity results in not finding the
primary criterion.

Since the process of scale-up is, by definition, an extrapolation,
it is important, after the preliminary scale-up criterion has been
defined, to do a rough design for the ultimate process scale.   If
any factors in the primary process criterion result in extrapolation,
it is good practice to return to the pilot plant to "stretch" the conditions to eliminate the extrapolation.   This might result in some ab-
normal experimental conditions.   An example of this occurs when

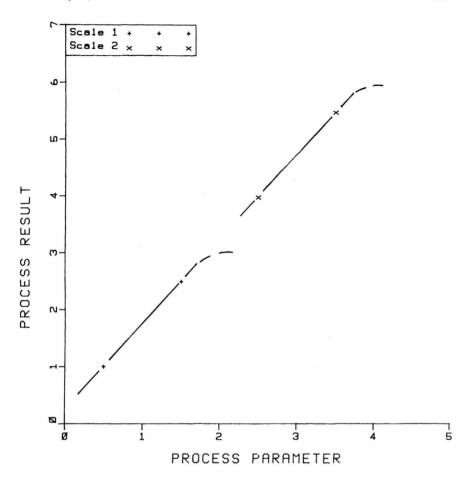

FIGURE 12   A case where a nonlinearity results in finding an in-correct primary criterion.

scaling-up a gas-liquid reactor; the superficial gas velocity for the full scale reactor would be much higher than would normally be en-countered in the pilot plant.

3.   *Choose the desired process result carefully.*   Improper choice of the desired process result can be the source of unexpected non-linearities.   This is especially true when dealing with subjective re-sponses such as flavor or texture.

Trying to gain some insight into the physics of the problem can help a great deal and often leads to choosing a process result which is more easily and more accurately measured. This can simplify interpretation of the experimental results and lead more readily to the correct choice of the primary scale-up criterion.

4. *Beware of minimal experimental designs.* Minimal experimental designs are insufficient for determination of scale-up criteria. Minimal experimental designs depend on the orthogonality of the process variables in order to be valid. When attempting to establish the primary scale-up criterion, the experimental variables are not generally the variables which will be used in the final correlations. For example, in the gas-liquid mixing example the experimental variables were rpm and gas flow rate. The variables used in the final correlation were power per unit volume and superficial velocity.

Figures 13 and 14 illustrate how an orthogonal experimental design, which incorporates the gas flow rate and rpm, is converted to a nonorthogonal design when the variables are converted to a delivery number and power per unit volume. Obviously, it is impossible to create an experimental design which is orthogonal for all the primary scale-up criterion candidates with the minimum number of data points. This implies that a significant number of points in excess of the minimum is required.

5. *Concentrate the experimental effort on the scale where data is most readily obtained.* There is no need to obtain complete sets of data on both experimental scales. Get most of the data on the scale where the experiments are the easiest to perform (usually the smallest), and then run only enough points to satisfy the determination of the primary scale-up criterion. There is an additional reason for this recommendation. Some of the examples that follow illustrate that under certain circumstances, careful analysis of the data obtained from one scale is sufficient to specify the primary scale-up criterion. For these cases, the second scale of experimentation may be eliminated.

IV. CASES REQUIRING ONLY ONE SCALE OF EXPERIMENTATION

In addition to the fortuitous case where the experiments illustrate that there is only one possible primary scale-up criterion, there are cases where two scales of experimentation may not be necessary. Generally these are associated with processes which are inherently independent of scale, such as illustrated in Table 4, or when equipment suppliers have done a thorough job of developing scale-up criteria. However, great caution should be exercised when accepting the supplier's scale-up criteria. There are very good reasons for extreme caution.

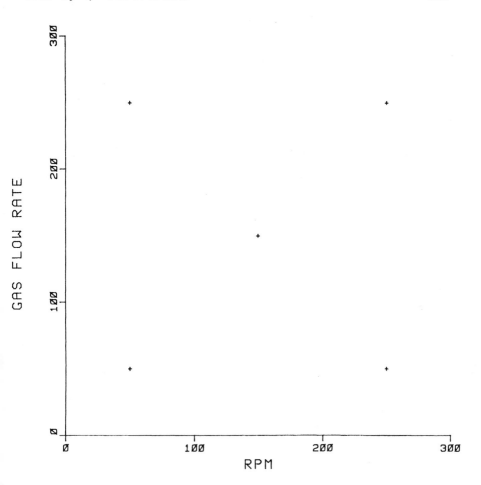

FIGURE 13   An orthogonal experimental design.

First, many suppliers have neither the time or resources to embark upon the complicated experimentation required to scale up. As a result, their "scale-up criteria" are often based upon a limited physical understanding of your specific process.

Further, even if the supplier has performed detailed scale-up experimentation, generally it will not have been performed on the exact product in question. Remember that scale-up must not result in any change in product performance and quality as perceived by the consumer. If a change occurs, it can have disastrous results.

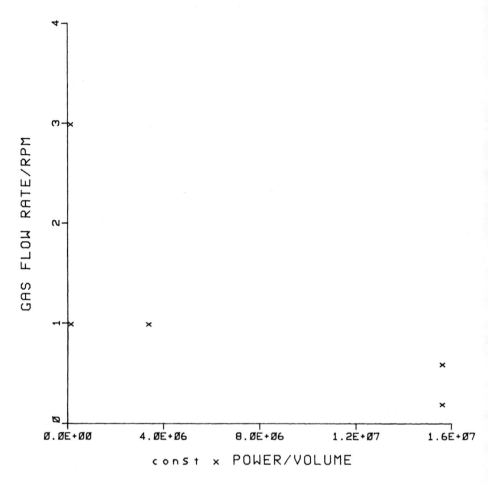

FIGURE 14   A design that has become nonorthogonal after data transformation.

Two of the examples that follow illustrate that a small change in product formulation can result in a complete change in the primary scale-up criterion.

It is desirable that scale-up be viewed as a cooperative effort between the supplier and purchaser. At the preliminary stages of process and product development, supplier's scale-up recommendations should be used as initial guidelines. Once experimental work on scale-up criterion is started, the suppliers' laboratory or pilot plant provides an excellent site for the second scale of experimentation.

There are four other cases when two scales of experimentation are not necessary:

1. *No actual scale-up is being done.* This occurs when the scaled-up process is actually an execution of the smaller scale in parallel. This does not mean simply constructing multiple copies of the identical pilot plant process, though this is one method of parallel scale-up. Another method of executing the smaller scale in parallel occurs when the process' physics is essentially two dimensional. Examples of this are the sheeting of doughs or drum drying of solids. In these cases, if the pilot scale of the process is of sufficient width to neglect edge effects, the process can be scaled-up by keeping a constant roll diameter and simply widening the rolls. The capacity of the process in these cases is simply proportional to the width of the rolls.

2. *If the process can be described through the use of dimensional analysis,* and all the dimensionless variables which describe the process can be maintained at their pilot plant levels upon scale-up, then complete similitude has been obtained, and no second scale of experimentation is required. This will be illustrated when dimensional analysis is discussed later in a following chapter.

3. *When the physics are well understood* and the desired process result can be included in a dimensionless equation obtained by a combination of dimensional analysis and experimentation, only one scale of experimentation is required. The cooking of the turkey is an example of this. Or, in cases when a fundamental or objective process result is desired, this technique is sufficient. The standard pressure drop and heat transfer correlations are examples of this technique.

4. *There are certain cases when all the potential primary scale-up criteria may only be varied in a simple manner on scale-up.* In these, very rare, cases, no scale-up experimentation is required. This is illustrated for the simple case of dispersion of gums later in this chapter.

V. EXTRUSION SCALE-UP

A. Primary Scale-Up Criterion for Extrusion Processes

Food extrusion is an area which has been frequently discussed as a difficult scale-up problem. Unfortunately there is little published data comparing two extruder sizes to illustrate the correct scale-up method.

The main difficulty in proper scale-up of extruders is not the estimation of the power and capacity. These are, in fact secondary criteria, and discussed frequently in the existing literature (7–10, 22–24,28–33,35–38,40,48,52–54,56,57,61,62,65–67). The most difficult problem encountered in scaling-up extrusion operations is in obtaining the same product qualities on the large scale extruder as achieved on the pilot plant or laboratory extruder. It follows that some measure of the product quality is the desired process result, and the primary design criterion(s) must ensure that the desired process result is maintained when the process is scaled up.

The mechanisms that control the ultimate quality of an extruded product are not well identified or quantified. As a result, scale-up of this process represents a significant challenge to the process development and design engineer. Consider the process as consisting of two distinct operations:

A transport (screw) and change of state (melting) process
A forming process

The first operation concerns itself with the physics occurring in the screw, the second with the forming die. Our intuition tells us that one, or both, of these operations, must have key effects on the ultimate quality of the product produced. Hypothesize various parameters which might be important. Some candidates are considered in Table 8.

TABLE 8 Primary Criterion Candidates for Extrusion Problems

Screw
  Shear rate developed
  Shear stress developed
  Specific mechanical energy input
  Specific thermal energy input
  Weighted average total shear strain
  Residence time
  A time-temperature integral representing extent of a reaction
    or physical process
  Product temperature
  Pressure changes
  Any combination of the above

Die
  Shear rate
  Shear stress
  Pressure drop
  Extrudate temperature

The reader can probably think of additions to this list. All the parameters listed are not necessarily independent of each other, but this will not be a concern at this time. Note that most of the variables are also strong functions of product formulation, which can have a significant effect on extrudate viscosity.

Assuming the desired process result has somehow been established, it will simplify the problem to assume that the product formulation is not varied. This is reasonable since changing formulae would result in a change in extrudate properties and product quality. Using a similar argument, assume that extrudate temperature is fixed, without regard to how this is to be accomplished. Using these assumptions, examination of the independent variables and parameters that (may) have an effect on the extrudate can occur. Small changes in moisture fed to the extruder might be conisidered in order to compensate for the normal variations that are observed in the "dry" raw materials or to explore the importance of these variations.

For the screw.

If the extruders being considered are geometrically similar, as illustrated in Figure 15, shear rate and shear stress are equivalent, because the definition of viscosity explicitly relates shear rate to shear stress by

$$\mu = \frac{\tau}{\dot{\gamma}} \tag{25}$$

where τ = shear stress, $\dot{\gamma}$ = shear rate, and μ = viscosity coefficient. Both are controlled by the screw speed and, to a lesser degree, the discharge pressure of the screw. For a specified screw speed and extruder feed rate, the pressure is completely controlled by the geometry of the die and the total number of die openings.

Specific mechanical energy is controlled by the die pressure and screw speed and by the feed rate to the screw. The feed rate is an independent variable only for screws which are fed in a "starved" condition. This is usually true for twin screw extruders.

All the other parameter candidates are controlled by the same variables given for shear rate and specific energy input.

Note that once extrudate temperature is fixed, thermal energy input is no longer an independent variable, since it is specified by specific mechanical energy input and the fixed extrudate temperature. Similarly, the temperature profile is unknown but specified.

The variables to be considered for describing the die are even more limited. Assume that the geometry of the die opening is

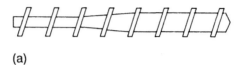

(a)

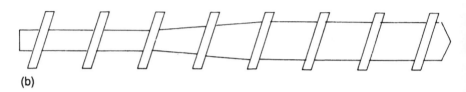

(b)

FIGURE 15   Geometrically similar extrusion screws.

fixed; the only variable effecting shear rate, shear stress, and
pressure drop, in a way that is independent of the screw's behavior, is the flowrate per die opening.   This variable is directly
manipulatible by opening or plugging die inserts.   The assumption
that the die geometry is not to be changed is very reasonable since
the final product dimensions are usually specified within a narrow
range.   Since all the parameter candidates are completely correlated
with one another, this type of experimentation does not yield any
insight into what the underlying physics of die/product interaction
behavior might be.   For example, correlation of the process result
with the die shear stress might indicate a mechanical failure of some
key component.   Fortunately this knowledge is not required to successfully scale-up the process.   Further understanding of the underlying phenomena could be obtained by running the experimental
design with a different die geometry.

To summarize, any experimentation to determine the primary
scale-up criterion for an extrusion operation, assuming that formula and extrudate temperature are fixed, has only relatively few
potential experimental variables.

Number of die openings
Screw speed
Feed rate, if the screw is starve fed

Table 9 provides first approximations for calculating the unmeasurable candidates for the primary scale-up criterion.

TABLE 9 Approximations for Unmeasurable Primary
Criteria Candidates

Shear rate developed
A very crude approximation is $\dot{\gamma} = \pi ND/b$

If rheological data is available, a better estimate can be obtained
from $\dot{\gamma} = (P_0/m\pi^1 ND^2 L)^{1/n}$

Shear stress developed
If rheological data is available, $\tau = m\dot{\gamma}^n$

Total shear strain
A crude approximation is $\bar{\gamma} = \dot{\gamma}\bar{t}$

Integral time/temperature history of a physical or chemical reaction
A crude approximation is $\psi = T_{avg}\bar{t}$

If a temperature profile and an activation energy are available, a

better approximation is $\psi = A \int_0^{\bar{t}} e^{-(\Delta EVRT)} \, dt$

For simple screw configurations, more exact formulas, found in
the literature (23,57) could be used. For complicated screw configurations, such as are normally found in twin-screw extruders,
exact calculation may be impossible. In any case, given the level
of uncertainty usually encountered in the actual experiments, the
effort expended in performing difficult calculations may not be
warranted.

Note that all the parameter candidates describing screw behavior
are independent of one another, but all are strong functions of the
key variable, screw speed, and extruder geometry. The fact that
screw speed plays such a prominent role in all the parameter candidates is the justification for using at least two scales of experimentation to determine the primary scale-up criterion. If only one
scale of experimentation is evaluated, the process result may be
correlated with any of the variables since all the variables are,
for a single scale, only transformations of screw speed. Any difference in the "quality of fit" obtained for regressions between a
chosen parameter and the desired process result is only the result
of the idiosyncracies of the regression techniques used. Unfortunately, this fact is not often recognized.

TABLE 10   A Hypothetical Process Result at Different
Screw Speeds[a]

Screw speed	Result	Screw speed	Result
5	28.40	55	64.92
10	5.53	60	72.77
15	6.49	65	85.97
20	5.38	70	76.85
25	25.62	75	62.70
30	16.88	80	58.30
35	11.30	85	95.95
40	62.73	90	114.30
45	51.55	95	117.80
50	58.92	100	102.33

[a]This table was generated by adding a random number
to a linear transformation of the screw speed.

The problem with experimentation at a single scale is supported
by an example. Table 10 presents a hypothetical data set illustrating a process result versus screw speed. For the purposes of
illustration, assume that two parameters being considered as candidates for the primary scale-up criterion are screw shear rate and
specific mechanical energy input. As a first approximation, shear
rate can be assumed directly proportional to screw rpm. If the
material is non-Newtonian, with a flow index of 0.5, the specific
mechanical energy input is, for a choke fed screw, roughly proportional to the 0.5 power of screw rpm (23,35,52).

If the data in Table 10 is reduced through the use of linear regression with rpm as the dependent variable the result is a correlation coefficient of 0.82, whereas the correlation using $(rpm)^{0.5}$ as
the dependent variable yields a correlation coefficient of 0.69. Does
the difference in the correlation coefficients provide a basis for selecting the primary scale-up criteria? The answer is no! In fact,
changing the method of regression using the logarithms of the two
candidates as the independent variables versus process result, identical correlation coefficients are obtained for both candidates.

B.   An Example for Twin-Screw Extruders

Data for the scale-up of a twin-screw cooking extruder are available in the literature and are useful for illustrating the scaleup

technique (47). The extrudate was potato starch and the twin-screw extruder was operated in a starved state. Several desired process results were considered including the percent solubility of the product, sedimentation volume of the product, and viscosity of the solubilized starch. Four primary variables were explored: screw speed, thermal energy input, feed rate, and extrudate moisture. Only the data for starch solubility will be discussed and are illustrated in Figures 16 and 17.

Figure 16 illustrates the relationship between solubility and extrudate temperature irrespective of specific mechanical energy input, and Figure 17 illustrates the relationship between solubility and specific mechanical energy input irrespective of extrudate temperature. Regression analysis resulted in the following equation:

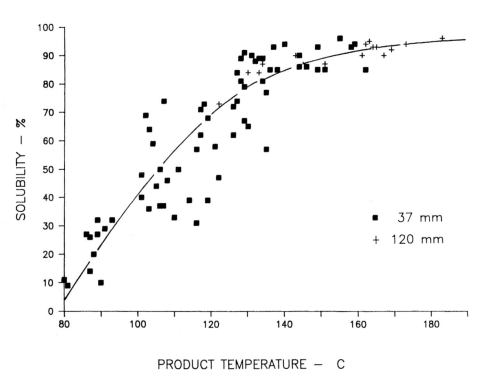

PRODUCT TEMPERATURE — C

FIGURE 16 Starch solubility versus extrudate temperature. (Adapted from Ref. 47.)

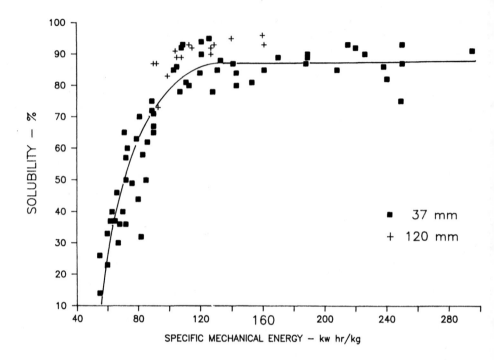

FIGURE 17   Starch solubility versus specific mechanical energy.
(Adapted from Ref. 47.)

$$\% \text{ Solubles} = -153.26 + 1.1525 \cdot \text{SME} + 1.804 \cdot \text{TP}$$
$$= -1.9686 \cdot 10^{-3} \cdot \text{SME}^2 - 3.653 \cdot 10^{-3} \cdot \text{TP}^2$$
$$= 3.178 \cdot 10^{-3} \cdot \text{TP} \cdot \text{SME} \qquad (26)$$

where

SME = specific mechanical energy input, kw hr/kg
TP = extrudate temperature, °C

There is difficulty with the methodology used for analysis.   The
candidates chosen as primary scale-up criteria parameters, product
temperature and specific mechanical energy input, are highly correlated with one another, calling into question the validity of the
regression result.

An alternative approach that avoids this difficulty is to consider specific mechanical energy input and specific thermal energy input as independent parameters. The final product temperature would then become a dependent variable. The use of composition (moisture) as an additional variable is not recommended, but did not seem to complicate the analysis. The point illustrated by Figures 16 and 17 and Eq. 26 is that the process result, for both scales studied, is defined by a combination of specific mechanical energy input and extrudate temperature.

As an additional point of interest, recent data (13) suggest very similar conclusions about the scale-up of a cornstarch cooking process. This work indicates the total specific energy input (mechanical + thermal) is the determinant of starch solubility and intrinsic viscosity. These authors have avoided the difficulty associated with using extrudate temperature as part of the primary criteria.

Note that any of the other primary criterion candidates, listed in Table 8, are mathematically independent of specific mechanical energy. As a result, if any of these were chosen, the data for the two scales would not have superimposed. The raw data has not been published, so the graphs for other candidates cannot be illustrated.

Additional publications (14,16,34,41—46,62—65) indicate that the suppliers of the type of equipment used generally suggests specific mechanical energy input as the primary scale-up criterion for a variety of products.

C. A Single-Screw Cooking Extruder Example

The data and examples that follow are being presented through the courtesy of The Pillsbury Company, Minneapolis, Minnesota. The data are presented in such a fashion as to protect proprietary information and yet provide very useful illustrations.

Consider a single screw cooking extruder processing a wheat flour-based product. The extruder was choke fed and externally heated through the use of a steam jacket. Throughout the experiment, discharge temperature was maintained within a very narrow range via manipulation of the steam pressure used in the jacket while keeping the composition of the extrudate fixed.

There are only two independent variables in this problem; the number of open die holes and the extruder speed (rpm). The desired process results considered were an overall hedonic grade, a firmness grade, and a product rehydration rate. The first two results were generated by a trained organoleptic panel.

The experiments were run in a very simple manner. A die geometry was selected and holder fabricated to hold various numbers

of die inserts. The process was modified by either changing the
speed of the extruder, and/or plugging inserts. In this manner,
a wide range of combinations of screw parameter candidates (Table
8) and die pressures could be obtained while keeping the two variables independent of one another.

For each experimental condition, the product was evaluated,
power consumption of the screw recorded, and the production rate
measured. Figures 18 through 20 are plots of the various desired
process results as a function of flowrate per die, with screw rpm
as a parameter. As explained above, flowrate per die defines the
only possible criterion candidate which describes the die, provided
that the composition, temperature, and geometry are fixed.

Consider Figure 18, the plot of overall hedonic grade versus
flowrate per die opening. Statistical analysis reveals that the inclusion of screw speed will not significantly improve the fit of the
data beyond that obtained by using flowrate per die opening alone.

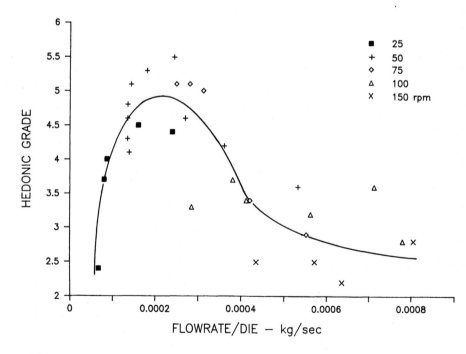

FIGURE 18 Hedonic grade versus die flow rate on single screw
cooking extruder.

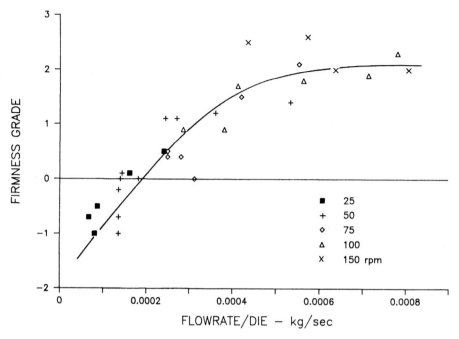

FIGURE 19   Firmness grade versus die flow rate on single screw cooking extruder.

In this case, the conclusion is that the design of the screw, or the conditions therein have little effect on the finished product quality. In a sense, there is no scale-up required, since the die geometry does not change as the process capacity increases.   The only change is the number of die openings required.   This is not to say that additional scale-up work is not required; a great many secondary criteria still need to be defined.   The required secondary criteria for extruders will be discussed later in this chapter.

The same conclusions can be drawn from Figures 19 and 20 which use firmness grade and rehydration rate as the desired process result.   In fact, further analysis revealed that the two "subjective" results, overall hedonic grade and firmness grade, are just measurements of one variable: rehydration.   The differences between Figure 18 and Figures 19 and 20 illustrate two points about experimental design that have not been discussed: The need for a careful choice of the desired process result, and the need for caution when dealing with nonlinear, subjective evaluations.   If the data used to obtain Figure 18 were gathered so that there was no data

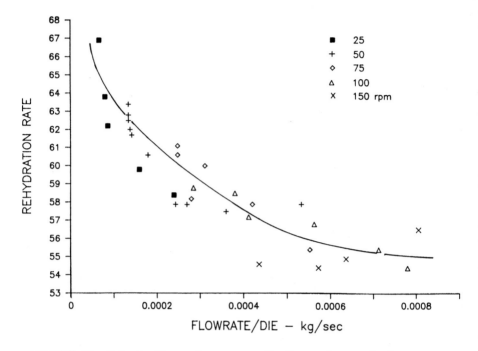

FIGURE 20   Rehydration rate versus die flow rate on single screw cooking extruder.

taken in the range of flowrate per die opening between 0.0002 kg/ sec and 0.0004 kg/sec, the correct scale-up criterion might not have been found!   This is illustrated in Figure 21.

D.   Two Single-Screw Forming Extruder Examples

Single-screw forming extruders have found wide application within the food industry, yet, with the exception of secondary criteria such as power, output, and heat transfer, little appears in the literature about the correct methods of scale-up.

The next examples illustrate an important point: The primary scale-up criterion is highly product dependent.   The two wheat-based products of almost identical composition were produced on virtually identical extruders, at the same extrudate temperature. The only difference was that one formula had a small percentage (approximately 0.5%) of inorganic salts that were added to give the product a unique functionality.

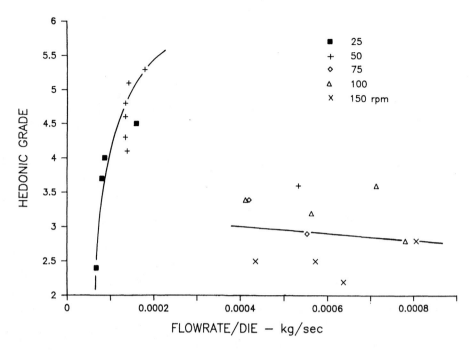

FIGURE 21 Misinterpretation as a result of missing data.

The experiments were performed in the same manner described for the above single-screw cooking extruder, except in this case, the temperature of the water used in the cooling jacket was manipulated to maintain the extrudate temperature within a narrow range (50 ± 2°C). The product moisture was allowed to vary over a narrowly prescribed range (30 ± 2%). Since the rheology of the two extrudates was well defined, this allowed for a more complete calculation of the scale-up criterion candidates. The product properties measured after the products were dried were rehydration of the salt-free product and density of the product containing the salt. The density measurement was found to be directly related to rehydration.

Figure 22 illustrates the result of scale-up experimentation for the salt-free product, and Figure 23 for the product containing salt over a range of number of die openings and screw speeds. The difference in the conclusions that can be drawn from the two studies is quite dramatic. For the product that did not contain

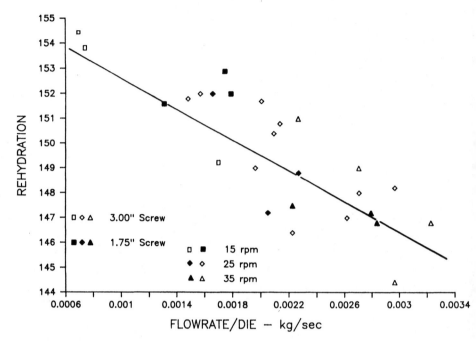

FIGURE 22   Rehydration versus die flow rate for a salt-free product on a forming extruder.

inorganic salts, die shear rate clearly defines the product quality independent of screw speeds and two extruder diameters. This is similar to the conclusion drawn previously for the single screw cooking extruder. In fact a second scale of experimentation was not necessary in these cases. For the product which contained the inorganic salts no relationship between die shear conditions was found. Rather, the two scales of extruder are described by a superficial screw shear rate.

The importance of the difference between these two conclusions cannot be overlooked. Clearly, the primary scale-up criterion is highly dependent on the composition and/or the required functionality of the finished product. This emphasizes the fact that some scale-up experimentation is required for almost any new product. Trials should always be run with the product in question on the supplier's equipment before one accepts their scale-up criteria.

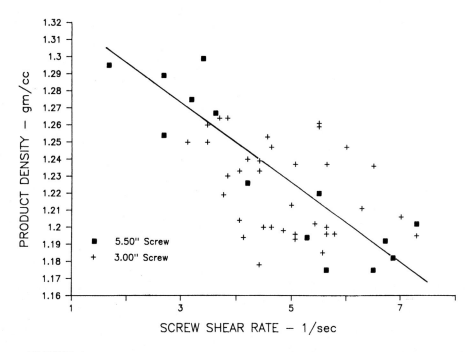

FIGURE 23 Density versus screw shear rate for a salt-containing product on a forming extruder.

E. Extrusion Secondary Scale-Up Criterion

In the previous examples, secondary criteria have been alluded to but not discussed. They cannot be overlooked. Once the primary scale-up criterion has been established, design of a full-scale system can proceed. A complete design cannot be executed without a complete knowledge of the secondary criteria.

For any extrusion scale-up, the following secondary criteria must be established:

The relationship between power consumption, screw speed, screw geometry, filled length, and screw backpressure
The relationship between die pressure drop and flow rate
For a choke fed extruder, the relationship between output and screw speed, screw geometry, and screw backpressure
For a starved screw, the relationship between filled length and screw geometry and feed rate

The relationship between heat transfer coefficients and extrusion conditions in order to design the heating/cooling system and to estimate wall temperatures, etc.
The likely formula variations the extruder will experience and the effect of these variations on extruder performance. Moisture variation of the feed materials is probably the most significant of the possible formula variations.

This list is not all-inclusive but contains the most significant secondary criteria.
Determination of the secondary criteria can be approached in a number of ways. In some cases there is sufficient literature to define the secondary criteria. This is particularly true for the behavior of single screw extruders. In many cases, the secondary criteria need to be experimentally determined. The best approach is through the use of deterministic modeling or dimensional analysis. In either case, some quantity of experimentation would be required to establish model parameters. In the next chapter, the use of dimensional analysis for the determination of secondary criteria for extrusion processes will be discussed in detail.

F. Reduction of Solids on Passing Through Blending Equipment

A common processing step is the passage of solid material through mixing equipment. One side effect of this process may be the desirable, or undesirable, reduction of some of the fragile solid particles present in the mixture. This can be a scale-up problem, since the mixers were probably not designed with blending or coating in mind, and not with particle size reduction. Figure 24, presented with the permission of the Pillsbury Company, Minneapolis, Minnesota, illustrates the primary scale-up criterion for the reduction of sugar through a commercially available high speed paddle mixer is power input per unit throughput of mix. These experiments were performed on a commercial grade of granulated sugar. The two experimental variables were mixer speed and feed rate. This is consistent with the normal scale-up criterion for devices designed for the grinding of solids (15,18,30,39).

G. An Example of Applied "Common Sense"

In some instances scale-up is nothing more than the application of "common sense" to a processing problem. Consider the process illustrated in Figure 25. This is commonly encountered in processes that require the dispersion of a solid in liquid media. The solid is often a difficult to disperse material, such as a vegetable gum or a starch. The dispersion device commonly encountered is essentially

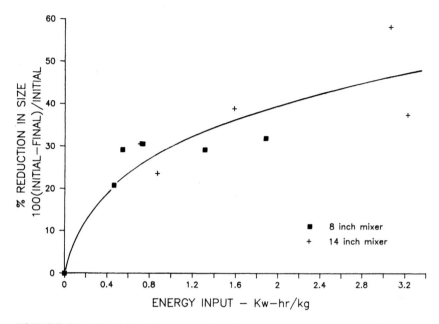

FIGURE 24 Particle size reduction in high-speed paddle mixers.

a horizontal centrifugal pump with a screen placed across the discharge.[2] The slight vacuum formed by the vortex in the blender tends to pull the solids into the blender. These devices are available in a variety of sizes, but the scale-up is not well understood. A common procedure is to test the full scale unit. As a consequence the only scale-up that is actually occurring is an increase in the size of the batch to be processed by increasing the size of the recirculation tank.

The desired process result of these devices is to disperse the solids in the liquid in such a way that the formation of agglomerates is avoided. For the case of dispersing gums, these agglomerates, called "fish eyes" because of their appearance, are very difficult, if not impossible, to break up or dissolve. As a result, the formation of "fish eyes" results in a reduced functionality of the

---

[2] These devices are manufactured by Tri-Clover, Inc., Kenosha, Wisconsin, under the trade name Tri-Blender.

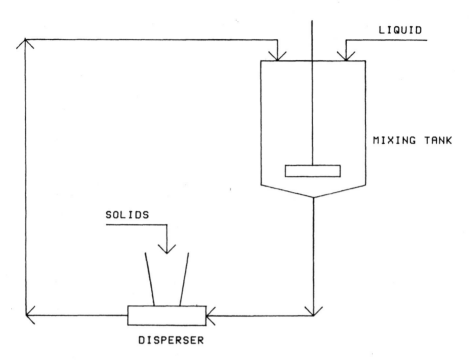

FIGURE 25   A process for dispersing solids in a liquid.

additives, which is usually measurable, as a reduced viscosity of
the finished product.

If one considers the criteria which might define the degree of
breakdown of the agglomerates as they pass through the high shear
region of the blender, several candidates emerge:

By analogy to grinding, the energy input per pound of solids might
   be indicative of particle reduction.
By analogy to liquid-liquid or liquid-gas mixing, the scale of turbulence or power per unit volume within the blender might be
   the determinate of particle breakdown.
The particles require a critical shear stress in order to "fail"—an
   analogy to the failure of ductile materials.
The total shear strain the product sees determines the elongation
   and failure of the particles.

With the constraint of fixed blender size, there is only one parameter available to maintain during scale-up to a larger recirculation

tank. Note that the energy dissipation rate (power/unit volume), shear rate, residence time, and shear stress developed within the blender are uniquely defined. As a consequence, we can draw the following conclusions:

The energy input per pound is simply proportional to the duration of recirculation through the blender divided by the mass of material in the recirculation tank.

The average total shear strain seen by the mixture is proportional to the time of recirculation through the blender divided by the mass of material in the recirculation tank.

The probability that any element of fluid will see the critical shear stress, turbulence level in the blender is defined by the number of turnovers in the tank.

In fact, the first two statements about energy input per pound and total shear strain are the same as the third. All specify a number of tank turnovers. The scale-up is therefore straightforward: Maintain the number of tank turnovers on scale-up, or the recirculation time is proportional to the mass in the recirculation tank (size of batch). This seems obvious but is often overlooked. Generally, the operator will turn off the recirculation pumps as soon as the addition of solids is complete. The natural tendency then is for the plant scale system to have fewer turnovers than the pilot scale system with poor dispersion of solids at the plant scale.

REFERENCES

1. Aiba, S., A. E. Humphrey, and N. Millis, *Biochemical Engineering*, Academic Press, New York (1973).

2. Bader, F. G., Improvements in multi-turbine mass transfer models, in *Biotechnology Progress*, C. S. Ho and J. Y. Oldshue, eds., American Institute of Chemical Engineers, New York (1987), pp. 96—106.

3. Balmer, G. J., I. P. T. Moore, and A. W. Nienow, Aerated and unaerated power and mass transfer characteristics of prochem agitators, in *Biotechnology Progress*, C. S. Ho and J. Y. Oldshue, eds., American Institute of Chemical Engineers, New York (1987), pp. 116—127.

4. Bird, R. B., W. E. Stewart, and E. N. Lightfoot, *Transport Phenomenon*, John Wiley and Sons, New York (1960).

5. Bondy, F. and S. Lippa, Heat transfer in agitated vessels, *Chem. Eng.* April:62—71 (1983).

6. Bridgeman, P. W., *Dimensional Analysis*, Yale University Press, New Haven (1922).

7. Carley, J. F., R. S. Mallouk, and J. M. McKelvey, Simplified flow theory for screw extruders, *Industrial and Engineering Chemistry*, 45(5):974–977 (1953).

8. Carley, J. F., and R. A. Strub, Application of theory to design of screw extruders, *Industrial and Engineerng Chemistry*, 45(5):978–988 (1953).

9. Carley, J. F., and R. A. Strub, Basic concepts of extrusion, *Industrial and Engineering Chemistry*, 45(5):970–973 (1953).

10. Carley, J. F., and J. M. McKelvey, Extruder scale-up theory and experiments, *Industrial and Engineering Chemistry*, 45(5): 989–991 (1953).

11. Carslaw, H. S., and J. C. Jaeger, *Conduction of Heat in Solids*, Clarendon Press, Oxford (1986).

12. Charm, S. E., *Fundamentals of Food Engineering*, AVI, Westport, Connecticut (1981).

13. Della Valle, G. Kozlowski, P. Colonna, and J. Tayeb, Starch transformation estimated by the energy balance on a twin screw extruder. *Lebensm. Wiss. Technol.*, in press (1989).

14. Dreiblatt, A., Accuracy in extruder scale-up, presented at the National Meeting of the American Institute of Chemical Engineers, August 17–19, Minneapolis, Minnesota (1987).

15. Earle, R. L., *Unit Operation in Food Processing*, Pergamon Press, Oxford (1983).

16. Eise, K., H. Herrmann, H. Werner, and U. Burkhardt, An analysis of twin-screw extruder mechanisms, *Advances in Plastic Technology*, 1(2):1–22 (1981).

17. Eliezer, E. D., Power absorption by new and hybrid mixing systems under gassed and ungassed conditions, in *Biotechnology Progress*, C. S. Ho and J. Y. Oldshue, eds., American Institute of Chemical Engineers, New York (1987), pp. 22–30.

18. Faust, A. S., L. A. Wenzel, C. W. Clump, L. Maus, and L. B. Andersen, *Principles of Unit Operations*, John Wiley and Sons, New York (1960).

19. Gbewonyo, K., D. DiMasi, and B. C. Buckland, Characterization of oxygen transfer and power absorption of hydrofoil impellers in viscous mycelial fermentations, in *Biotechnology Progress*, C. S. Ho and J. Y. Oldshue, eds., American Institute of Chemical Engineers, New York (1987), pp. 128–134.

20. Geankoplis, C. J., *Transport Processes and Unit Operations*, Allyn and Bacon, Boston (1983).

21. Hayt, H. H., and J. E. Kimmerly, *Engineering Circuit Analysis*, McGraw-Hill, New York (1978).

22. Harper, J. M., and D. V. Harmann, Effect of extruder geometry on torque and flow, *Transactions of the ASAE*, 16(6): 1175–1178 (1973).

23. Harper, J. M., *Extrusion of Foods*, CRC Press, Boca Raton, Florida (1981).

24. Harper, J. M., Food extrusion, *CRC Critical Review in Food Science and Nutrition*, 11(2);155–215 (1979).

25. Heldman, D. R., and R. P. Singh, *Food Process Engineering*, AVI, Westport, Connecticut (1981).

26. Ho, C. S., and M. J. Stalker, The oxygen transfer coefficient in aerated stirred reactors and its correlation with oxygen diffusion coefficients, in *Biotechnology Progress*, C. S. Ho and J. Y. Oldshue, eds., American Institute of Chemical Engineers, New York (1987), pp. 85–95.

27. Isaacson, E., and M. Isaacson, *Dimensional Methods in Engineering and Physics*, John Wiley and Sons, New York (1975).

28. Janssen, L. P. B., *Twin Screw Extrusion*, Elsevier Science Publishing Co., New York (1978).

29. Jepson, C. H., Future extrusion studies, *Industrial and Engineering Chemistry*, 45(5):992–993 (1953).

30. Johnston, R. E., and M. W. Thring, *Pilot Plant Models and Scale-Up Methods in Chemical Engineering*, McGraw-Hill Book Company Inc., New York (1957).

31. Klein, I., Extrusion—an update on melt technology, part one, *Plastics Design and Processing*, August:10–15 (1973).

32. Klein, I., Extrusion—an update on melt technology, part two, *Plastics Design and Processing*, September:22–24 (1973).

33. Klein, I., Extrusion—an update on melt technology, part three, *Plastics Design and Processing*, October:20–23 (1973).

34. Kuhle, R., Continuous dough manufacturing system, presented at the National Meeting of the American Institute of Chemical Engineers, November 17–19, Miami, Florida (1986).

35. Levine, L., Estimating output and power of food extruders, *Journal of Food Process Engineering*, 6:1–13 (1982).

36. Levine, L., and J. Rockwood, Simplified models for estimating isothermal operating characteristics of food extruders, *Biotechnology Progress*, 1(3):189 (1985).

37. Mallouk, R. S., and J. M. McKelvey, Power requirements of melt extruders, *Industrial and Engineering Chemistry*, 45(5): 987–988 (1953).

38. Martelli, F. G., *Twin Screw Extruders: A Basic Understanding*, Van Nostrand Reinhold Company, New York (1983).

39. McCabe, W. L., J. C. Smith, and P. Harriott, *Unit Operations of Chemical Engineering*, McGraw-Hill, New York (1985).

40. McKelvey, J. M., Experimental studies of melt extrusion, *Industrial and Engineering Chemistry*, 45(5):982–986 (1953).

41. Meuser, F., and B. Van Lengerich, Possibilities of quality optimization of industrially extruded flat breads, in *Thermal Processing and Quality of Foods*, P. Zeuthen, J. C. Cheftel, M. Jul,

H. Leniger, P. Linko, F. Varela, and G. Vos, eds., pp. 180–184, Elsevier Applied Science Publishers, New York (1984).

42. Meuser, F., and B. Van Lengerich, System analytical model for the extrusion of starches, in *Thermal Processing and Quality of Foods*, P. Zeuthen, J. C. Cheftel, M. Jule, H. Leniger, P. Linko, F. Varela, and G. Vos, eds., pp. 175–179, Elsevier Applied Science Publishers, New York (1984).

43. Meuser, F., B. Van Lengerich, and E. Groneick, The use of high temperature short time extrusion cooking of malt in beer production, in *Thermal Processing and Quality of Foods*, P. Zeuthen, J. C. Cheftel, M. Jul, H. Leniger, P. Linko, F. Varela, and G. Vos, eds., pp. 127–136, Elsevier Applied Science Publishers, New York (1984).

44. Meuser, F., B. Van Lengerich, and F. Kolher, Extrusion cooking of protein and dietary fiber enriched cereal products: nutritional aspects, personal correspondence with B. Van Lengerich, Werner Pfleidderer Corporation, Ramsey, New Jersey (1986).

45. Von Meuser, V., B. Lengerich, and F. Kohler, Einflub der Extrusion Extrusionsparameter auf funktionelle Eigenschaften von Weizenstarke, *Starch/Starke*, 34(11):366–372 (1982).

46. Meuser, F., B. Van Lengerich, A. E. Pfaller, and A. E. Harmuth-Hoene, The influence of HTST-extrusion cooking of the protein nutritional value of cereal based products, in *Extrusion Technology for the Food Industry, Part II, Aspects of Technology*, P. Colonna, ed., pp. 35–53, Elsevier Applied Science Publishers, New York (1987).

47. Von Meuser, V., B. Van Lengerich, and H. Rheimers, Kochextrusion von Starken, *Starch/Starke*, 36(6):194–199 (1984).

48. Middleman, S., *Fundamentals of Polymer Processing*, McGraw-Hill Book Co., New York (1981).

49. Mohesenin, N. M., *Thermal Properties of Food and Agricultural Materials*, Gordon and Breach, London (1980).

50. Nagata, S., *Mixing Principles and Applications*, Wiley, New York (1975).

51. Oldshue, J. Y., *Fluid Mixing Technology*, McGraw-Hill, New York (1983).

52. Schenkel, G., *Plastics Extrusion Technology and Theory*, American Elsevier Publishing Company, New York (1966).

53. Schopf, L., Scale-up of twin screw extruders, course notes for Food Extrusion, Center for Professional Advancement, New Brunswick, New Jersey (1986).

54. Stevens, M. J., *Extruder Principles and Operation*, Elsevier Applied Science Publisher, New York (1985).

55. Sears, F. W., and M. W. Zemansky, *University Physics, Part 2*, Addison Wesley, Addison, Massachusetts (1964).

56. Tadmor, Z., and C. G. Gogos, *Principles of Polymer Processing*, Wiley, New York (1979).

57. Tadmor, Z., and I. Klein, *Engineering Principles of Plasticating Extrusion*, Van Nostrand Reinhold Company, New York (1970).

58. Taylor, E. S., *Dimensional Analysis for Engineers*, Clarendon Press, Oxford (1974).

59. Uhl, V. W., and J. B. Gray, *Mixing Theory and Practice*, Academic Press, New York (1967).

60. Wang, D. C., C. L. Cooney, A. L. Demain, P. Dunnill, A. E. Humphrey, and M. D. Lilly, *Fermentation and Enzyme Technology*, Wiley, New York (1979).

61. Wilkinson, W. L., *Non-Newtonian Fluids*, Pergamon Press, New York (1960).

62. Yacu, W. A., Energy balance in twin screw co-rotating extruders, presented at the A.A.C.C. short course on extrusion, San Antonio, Texas (1987).

63. Yacu, W. A., Extrusion cooking analysis: I. Processing aspects of twin screw co-rotating extruders, presented at the American Association of Cereal Chemists' short course on extrusion, May 2–3, San Antonio, Texas (1987).

64. Yacu, W. A., Extrusion cooking analysis: II. Extrudate physical and functional properties, presented at the American Association of Cereal Chemists' short course on extrusion, May 2–3, San Antonio, Texas (1987).

65. Yacu, W. A., Modeling of a twin screw extruder, *Journal of Food Engineering*, 8:1–21 (1985).

66. Zamodits, H. J., and J. R. A. Peason, Flow of polymer melts in extruders, *Transactions of the Society of Rheology*, 13(3): 357–385 (1969).

67. Van Zuilichem, R., Similarity, scale-up, and thermal behavior of extruders, course notes for Food Extrusion, Center for Professional Advancement, New Brunswick, New Jersey (1986).

11
Dimensional Analysis

I. INTRODUCTION

Dimensional analysis is a useful and powerful tool for solving scale-up problems. Dimensional analysis is the method of reducing the equations that describe a process into a form containing no reference to units of measurement, that is, a dimensionless form.

By its very nature, dimensional analysis gives a relationship between the independent and dependent variables that is independent of scale. As a consequence, the analysis of a problem through dimensional analysis, if it is possible, will inevitably lead to a solution of the scale-up problem. At the very least, a dimensional analysis will lead to a clearer understanding of the problem.

Unfortunately, there is a large class of problems, associated with the scale-up of food processes, to which dimensional analysis cannot be applied: the analysis of the effect of process variables on most product qualities. This conclusion may be simply summarized by saying there is no succinct definition of a dimensionless variable for flavor or texture.[1] The literature (8,31,34,56), although it does

---

[1]This statement is not absolutely correct. There have been attempts to quantify texture measurements in such a way as to make them suitable for dimensional analysis (see, for example, Ref. 13). These efforts have really been confined to the analysis of physical properties of the food product being masticated, not the interaction of these properties with the process used to manufacture the product.

not discuss the type of properties with which the food technologist
normally deals, states the problem as the inability of dimensional
analysis to deal with attributes lacking the property of absolute
significance of relative magnitude. The most common example of this
is the use of the Celsius or Fahrenheit scales of temperature, instead of the absolute Kelvin or Rankine scales. There is no problem when dealing with temperature differences. A better example
of lacking the property of absolute relative magnitudes is the hard-
ness scales that are used to describe materials. The common Mohs
scale, which assigns a value of 10 to a diamond and a value of 1 to
talc, does not mean that diamonds are 10 times harder than talc particles. Similarly, a grade of 10 on an orgaoleptic scale does not imply that the product tastes 10 times better than a product that was
assigned a grade of one. All that can be said about these types of
scales is that the higher the grade, the better, or harder, the product.

Fortunately, there are a large class of scale-up problems which
do not refer to the subjective quality scales. The most prominent
of these are the secondary criteria associated with the process design. Dimensional analysis can almost invariably be applied to these
aspects of the scale-up problem. In addition there are classes of
primary criteria to which dimensional analysis can be applied. Two
examples of this are the uniformity resulting from the blending of
simple mixtures and "objective" processes, such as the cooking of
a turkey to a specified temperature endpoint.

Another area to which dimensional analysis can be applied, but
which is often overlooked, is the preliminary analysis of problems
where some of the underlying physics is understood. An example
of this would be qualitatively predicting whether scale-up of a process is likely to increase the stresses or shear that a product will
undergo. In a case such as the mixing or sheeting of wheat doughs,
it is well known that these stresses may have a deleterious effect on
product quality caused by damaging the gluten matrix formed during dough development. A dimensional analysis will immediately indicate whether or not the forces or stresses of interest are likely
to increase on scale-up. This warns of a likely scale-up problem.

To summarize the values of dimensional analysis:

A relationship is obtained beween the process independent and dependent variables that is independent of process scale. As a
   consequence, it may be possible to eliminate the second scale of
   experimentation in order to scale up a process.
Dimensional analysis simplifies the view of the problem, making data
   analysis more straightforward.
It may be used to gain insight into the nature of the scale-up problem, without experimentation.

Associated with these benefits are some limitations:

Dimensional analysis cannot be easily applied to many problems involving qualitative measurements related to organoleptic characteristics of a product.
A fair degree of physical insight is required to successfully execute a dimensional analysis.
It may not simplify the problem.

The last two limitations will be illustrated later in this chapter.

II. THE METHODS OF DIMENSIONAL ANALYSIS

The literature (4,8,23,30,31,34,36,56) describes many techniques for performing dimensional analysis. Although the techniques may vary in detail, they all fall into two basic categories:

Normalization and scaling of the underlying differential equations
Application of a Buckingham π theorem

The first technique is based on the fact that the dimensionless solution to any differential equation(s) must be a function only of the dimensionless variables and parameters that appear in the differential equation(s), and its associated boundary conditions, after the equation(s) and boundary conditions have been transformed into a dimensionless form.
The Buckingham π theorem may be stated as follows:

The relationship between N dimensional variables, containing M fundamental dimensions, may be represented through the use of N-M dimensionless groups.

Needless to say, the basis for this theorem is not intuitive or obvious. The proof may be found in several texts (8,31).
Each technique for dimensional analysis has advantages and disadvantages. In order to normalize and scale the underlying differential equations and boundary conditions, one must be able to write the fundamental equations. This requires a level of understanding of the physics that is not always available. Assuming that the differential equations and boundary conditions can be written, performing the analysis on the differential equations is always preferable to resorting to the Buckingham π theorem, which yields less information.
On the other hand, the Buckingham π theorem must be used when the underlying differential equations cannot be written, and the

technique is rapid and easily executed for a problem of any degree of complexity.

III. USING DIFFERENTIAL EQUATIONS

A. Normalization and Scaling of the Differential Equations

If one can write the differential equations that describe a process, dimensional analysis via normalization and scaling is a straightforward technique. The procedure is as follows:

1. Choose scaling factors for the dependent variables that, if possible, result in ranges of the scaled dimensionless variables of 0 to 1. Of course, certain variables have a range of infinity, such as time in the nonsteady state heat transfer equation. In that case, the scaled variable should range from 0 to ∞. The scaling factors should be made up of parameters that appear in the original differential equation.
2. Substitute the scale variable into the differential equation and boundary conditions.
3. Rearrange the differential equation and boundary conditions so that a minimum number of parameters are left.

Sometimes the natural scaling factors are apparent, sometimes they are not. After some practice, the best choice of scaling factors will be easy to identify.

The general technique of scaling a differential equation is illustrated in the following examples.

B. Dimensional Nonsteady State Heat Transfer to Solids

The differential equation for nonsteady state heat transfer to a sphere is (4,11)

$$\frac{\partial T}{\partial t} = \frac{\alpha}{r^2} \frac{\partial}{\partial r} \left(r^2 \frac{\partial T}{\partial r} \right) \tag{1}$$

where

t = the time
r = the distance in the radial direction
T = the temperature
α = the thermal diffusivity of the material (L^2/t)

Generally, the boundary conditions associated with this differential equation are:

An initial condition,

$$\text{at } t = 0, \quad T = T_0 \tag{1a}$$

Two boundary conditions: (1) Either

$$\text{at } r = 0, \quad \partial t / \partial r = 0 \tag{1b}$$

or

$$\text{at } r = 0, \quad T = \text{finite} \tag{1c}$$

(2) And either, the surface temperature is defined by

$$\text{at } r = R, \quad T = T_a \tag{1d}$$

where T_a = the ambient temperature; or the surface heat flux is defined through an equation describing convective heat transfer.

$$\text{at } r = R, \quad k(\partial T / \partial r) = h(T_a - T) \tag{1e}$$

where h = a surface heat transfer coefficient and k = the conductivity of the sphere.

The choices of the scaled variable for the radial distance and temperature are straightforward. Define

$$r^* = r/R \tag{2}$$

$$T^* = (T - T_a)/(T_0 - Ta) \tag{3}$$

Both of these transformations lead to a range of the scaled variables of 0 to 1.

Then

$$\partial r^* = \partial r / R \tag{2a}$$

$$\partial r^{*2} = \partial r^2 / R^2 \tag{2b}$$

$$\partial T^* = \partial T / (T_0 - T_a) \tag{3a}$$

$$\partial^2 T^* = \partial^2 T / (T_0 - T_a) \tag{3b}$$

The choice of the scaling factor for time is not so obvious. The only choice that can be reached through the use of the parameters in the original equation is

$$t^* = \alpha t / R^2 \tag{4}$$

which leads to

$$\partial t^* = \alpha \partial t / R^2 \tag{4a}$$

Substituting these variables into the differential equation yields

$$\alpha \frac{(T_a - T_0)}{R^2} \frac{\partial T^*}{\partial t^*} = \frac{\alpha}{R^2 r^{*2}} \left[\frac{1}{R} \frac{\partial}{\partial r^*} \left(\frac{R^2 r^*}{R} \frac{\partial^2 T^*}{\partial r^*} \right) \right] \tag{5}$$

Rearrangement yields

$$\frac{\partial T^*}{\partial t^*} = \frac{1}{r^{*2}} \left[\frac{\partial}{\partial r^*} \left(r^{*2} \frac{\partial T}{\partial r^*} \right) \right] \tag{6}$$

For the boundary condition that specifies a surface temperature, scaling the boundary condition adds no dimensionless groups. It follows that

$$T^* = f(\alpha t / R^2, r^{*2}) \tag{7}$$

The dimensionless term containing the time, thermal diffusivity, and radius appears frequently in analyses of heat transfer and is called the Fourier number.

A great deal of information has been gained without actually solving the differential equation. Consider the case of the "spherical" turkey. In that example, the time to cook the turkey to some specified center temperature at some specified oven temperature was sought. Since the oven and final center temperature are specified, the following are specified:

$$T^* = \text{constant} \tag{8a}$$

$$r^* = 0 \tag{8b}$$

Utilizing this information in Eq. 7 yields

$$\text{constant} = f(\alpha t / R^2) \tag{9}$$

which implies

$$\alpha t / R^2 = \text{constant} \tag{10}$$

This is the same answer found earlier by a much more circuitous route. Note that in this case the constant could have been determined with only one experiment. That constant would then describe the time required to cook a turkey of any size. Repeating this exercise for any coordinate system yields a general rule which may be stated as

The time required to heat or cool any regular object a specified degree is proportional to the square of the characteristic dimension of the object.

This rule will be used in later examples to extract additional information from more complicated dimensional analyses.

This simple exercise illustrates one of the great powers of dimensional analysis. The solution to a general class of problems has been formulated with minimum experimentation and minimum mathematical effort.

If the boundary condition (Eq. 1e) incorporating convective heat transfer were included in the analysis, another dimensionless group would appear from

$$\text{at } r^* = 1, \ -(k/R)(dT^*/dr^*) = h \tag{11}$$

Rearranging yields

$$(hR/k) = -(dT^*/dr^*)_{r^* = 1} \tag{12}$$

It follows that the solution to the differential equation takes the functional form:

$$T^* = g(\alpha t/R^2, r^*, hR/k) \tag{13}$$

The term containing the heat transfer coefficient, thermal conductivity, and radius is known as the Biot number. The forms of the functions in Eqs. 7 and 13 may be found in any number of literature references (4,11,13,23,29).

Equation 13 is not as readily useful as Eq. 7 since the major dimension appears in two dimensionless groups. This prevents the determination of the effect of object size by inspection. Nonetheless, the problem has been significantly simplified by reducing the solution to a function of just two variables: the Fourier number and the Biot number. A solution defined in terms of these true variables will be true not just for any scale, but for any material.

The problem may be extended to any arbitrary geometry without great difficulty. For a three-dimensional problem in cartesian coordinates, the heat transfer equation is

$$\frac{\partial T}{\partial t} = \alpha \left(\frac{\partial^2 T}{\partial x^2} + \frac{\partial^2 T}{\partial y^2} + \frac{\partial^2 T}{\partial z^2} \right) \tag{14}$$

For purposes of illustration, assume that the surface temperature of the object is specified and that it is isentropic. The scaling of the differential equation proceeds as before, except each direction must be scaled independently. It is not difficult to show that

$$T^* = g(\alpha t/X^2, x^*, y^*, z^*, X^2/Y^2, X^2/Z^2) \qquad (15)$$

where X, Y, Z = the significant dimensions in the x, y, z directions and X = the smallest dimension.

Equation 15 yields some very important information. If geometric similarity is maintained on scale-up, the terms containing the ratios of the square of the dimensions are constants, and Eq. 15 reduces to the same form as Eq. 7. We can conclude that the general rule about heating and cooling is true for all shapes, provided geometric similarity is maintained. Additionally, the terms containing the squares of the geometric ratios are the source of the form of the empirical shape correction factors (13,29,45) found in such heat transfer problems as the freezing of foods.

C. Dimensional Analysis of Flow Problems

The analysis of flow problems follows the same procedure as that used for heat transfer. Consider the laminar flow through a narrow slot. The differential equation for this situation is

$$\frac{\Delta P}{L} = \mu \frac{\partial^2 v}{\partial x^2} \qquad (16)$$

where

v = the fluid velocity
P = the pressure drop
L = the length of the slot
μ = the viscosity of the fluid
x = the direction normal to the flow direction

The appropriate boundary conditions are

at x = ±b/2, v = 0   (16a)

at x = 0, dv/dx = 0   (16b)

where b = the depth of the slot.

The first boundary condition is the assumption of no slip at the wall. The second takes symmetry into account. From the continuity equation,

$$Q = V_{avg}*b = \int_{-b/2}^{b/2} v \, dx \qquad (17)$$

where Q = the volumetric flowrate per unit width and V_{avg} = the average fluid velocity. The logical scale factors to use are

$$v^* = v/V_{avg} \tag{18a}$$

$$x^* = x/b \tag{18b}$$

Substituting into the differential (Eq. 16) and rearranging yields

$$v^* = f(Pb^2/L\mu V_{avg}, x^*) \tag{19}$$

This equation may be integrated (Eq. 17) once over the region over the height of the slot to obtain the total flow rate. This removes x^* from the result, and

$$Pb^2/L\mu V_{avg} = \text{constant} \tag{20}$$

This relates the pressure drop to the velocity and size of the slot, giving a proportionality constant. Algebraic manipulation results in a more familiar form:

$$(Pb/4L/\rho V_{avg}^2/2 = \text{constant}(\mu/\rho b V_{avg}) \tag{21}$$

The two new terms will be recognized as a friction factor and the reciprocal of a Reynolds number. The constant, which may be found in the literature (44), has a value of 24. Again the key idea is that most of the conclusions obtained from conventional, difficult analyses can be obtained quickly and easily through dimensional analysis.

In a manner similar to that used for the heat transfer problem, it can be shown that extension of the solution to arbitrary geometries results in the addition of terms containing the squares of geometric dimensions. As long as geometrically similar systems are considered,

The pressure drop per unit length for *laminar* flow, of a Newtonian fluid, is always directly proportional to the average velocity and viscosity, and is inversely proportional to the square of the characteristic dimension.

Admittedly the rule is limited to Newtonian fluids, which are not commonly encountered in the design of food processing systems. This is not a major problem, since the analysis can be extended to non-Newtonian flow situations which may be characterized as power law fluids, and can be used to characterize most fluid foods that will be encountered. The apparent viscosity of a power law fluid is given by

$$\tau = m\dot{\gamma}^n \tag{22}$$

where

τ = the shear stress
γ = the shear rate
n = the power law flow index
m = the flow consistency

It is left as an exercise for the reader to show that the equivalent of Eq. 20 for a power law fluid is

$$Pb^{n+1}/mLV_{avg}{}^n = \text{constant} \tag{23}$$

Equation 23 provides a rule describing most fluids that one is likely to encounter in food processing.

By following a similar procedure, a general rule for viscous power consumption can be developed for various systems. The general rule for power law fluids is

The power consumption for laminar flow is proportional to a characteristic dimension to the 2 - n power and the characteristic velocity to the 1 + n power.

This will be developed during the dimensional analysis of extruders.

D. Dimensional Analysis of the Behavior of Screw Extruders

Extrusion is a commercially important unit operation that is common to many segments of the food processing industry. For any extrusion scale-up, the engineer is concerned with the prediction of output, power, and heat transfer in a food extruder.

Consider a system where the screw is full and the properties of the extrudate are not a function of temperature. The consequences of these simplifications will be discussed later. To illustrate, consider the simplified parallel plate model of the single-screw extruder, shown in Figure 1, that is widely discussed in the literature (5,9, 27,28,38,40,44,51,52,54,55,57,59). If the extrudate is assumed to obey the power law model for fluid flow, which represents many food extrudates (12,27,28,33,37,38,47,50), the differential equation which describes the flow situation is

$$\left(\frac{\Delta P}{L}\right) = m \frac{\partial}{\partial z}\left[\left(\frac{\partial v_x}{\partial z}\right)\right]^n \tag{24}$$

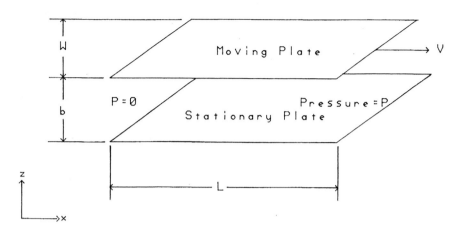

FIGURE 1   Two-plate approximation of flow in an extruder.

where

n = the power law flow index
m = the power law flow consistency
ΔP = pressure rise in extruder

Subject to the boundary conditions:

at z = 0, v = 0 (no slip)   (24a)

at z = b, v = V $\approx$ screw speed   (24b)

Choose the following scale factors:

$v^* = v/V$ (25a)

$z^* = z/b$ (25b)

Then

$dv^* = dv/V$ (25c)

$dz^* = dz/b$ (25d)

After substitution and rearrangement, the differential equation and boundary conditions become

$$\frac{\Delta P}{mL}\left(\frac{b^{n+1}}{V^n}\right) = \frac{\partial}{\partial z^*}\left(\left[\frac{\partial v^*}{\partial z^*}\right]^n\right) = \pi_1 \tag{26}$$

$$v^* = 0, \text{ at } z^* = 0 \tag{26a}$$

$$v^* = 1, \text{ at } z^* = 1 \tag{26b}$$

In addition to dimensionless velocity and position, there are two dimensionless parameters left in the differential equation (remember that n, the flow index, is a dimensionless parameter). It follows that the solution must take the following form:

$$V^* = F(x^*, \pi_1, n) \tag{27}$$

Since the designer is more interested in extruder output than local velocities, integrate across the channel cross section to obtain a relationship for extruder output. The result is a dimensionless expression for output as a function of one dimensionless parameter:

$$\frac{Q}{WVb} = g(\pi_1, n) \tag{28}$$

where Q = the volumetric output of the extruder.

This problem has been solved numerically. The results can be found in several sources (5,27,28,44,51,52,54,55). A graphical representation of the solution is presented in Figure 2.

The analysis that has been performed does not include any transverse flows and is missing additional details of the screw geometry. It is not difficult to show that the solution presented in Eq. 28 will be modified by the addition of terms containing geometric ratios, but will retain the same form. These geometric corrections can be found in the literature (10,27,28,44,51,52,54,55). The fact is that all the geometric corrections need not be known to use Eq. 28 to solve scale-up questions.

The basic differential equation, Eq. 24, if one includes all three dimensions, and all the possible three-dimensional flows, applies to any geometry, provided that the screw is operated in a filled condition and isothermally, or the extrudate properties are constant. This includes corotational or counterrotational twin screws, or variable geometry single screws, or any other configuration encountered. For more complicated situations, it may be impossible to solve the differential equations, or even to adequately define the boundary conditions. This is of no consequence. As long as one maintains geometric similarity of the cross section between the pilot plant and plant scales, Eq. 24 is applicable and can be easily used to design an extruder of any size. In this manner the secondary criteria governing the extruder output are readily determined.

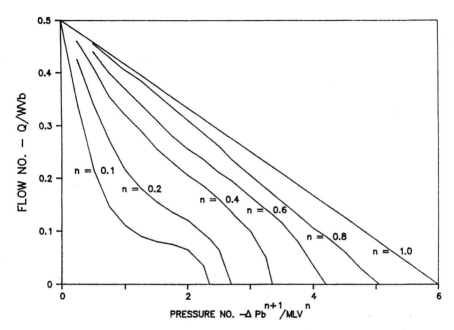

FIGURE 2 Dimensional relationship between extruder output and pressure. (From Ref. 44, used with permission.)

Constant Extrudate Properties

The technique for applying Eq. 28 is straightforward. One needs to experimentally generate the equivalent of Figure 2 for the extruder of interest. This plot would then be valid for the identical material on the larger-scale extruder having a geometrically similar cross section. Note that this requires only a partial determination of the rheology of the material: Only the flow index of the material has to be known, since the flow consistency may be carried through all the calculations as a proportionality constant. In fact, it will be demonstrated that the same experiments used to determine the dimensionless relationship for output and power will yield a working estimate for the flow index.

Next the scale-up of power consumption is considered. Power consumption is defined by

$$P_0 = FV \qquad (29)$$

where P_0 = power consumption; F = the drag force on the plate in Figure 1.
The drag force is simply the shear stress on the plate times the area of the plate. The shear stress at the plate is defined, through the use of the power law, by

$$\tau_{z=b} = m\left(\frac{\partial v_x}{\partial z}\right)^n_{z=b} \tag{30}$$

In principle, Eq. 27 can be differentiated to yield the shear rate in the fluid. The shear stress at the moving plate can then be calculated through the application of Eqs. 29 and 30. Performing this task yields, after some rearrangement, a dimensionless relationship for power.

$$\frac{P_0 b^n}{mV^{n+1}WL} = h(\pi_1, n) \tag{31}$$

Physical intuition can be used to simplify and extract additional information from the dimensionless equation for power (Eq. 31). If the problem is reworked with pressure rise in Eq. 26 having a value of zero, the unknown function in Eq. 31 assumes a value of unity. This allows the following modification to Eq. 31.

$$\frac{P_0 b^n}{mV^{n+1}WL} = 1 + \tilde{h}(\pi_1, n) \tag{32}$$

For most food extrusions, the power dissipated in moving the screw is much greater than the power consumed in creating the discharge pressure. Readers can readily convince themselves of this fact by comparing the total horsepower of the extruder to the product of volumetric screw output and pressure. This allows the conversion of Eq. 32 to the following approximate relationship:

$$\frac{P_0}{\rho V^3 WL} = \frac{m}{\rho V^{2-n} b^n} \tag{33}$$

This now fits the general rule, stated earlier, for predicting the relationship between power consumption, scale, and speed. The dimensionless term P_0/V^3WL is commonly known as a power number or the Lagrange group. The dimensionless term on the right-hand side of the equation is the reciprocal of a Reynolds number.

Equation 33 can be used to obtain a working approximation for the flow index n of the extrudate. The equation can be rearranged to the form $P_0 \alpha V^{1+n}$ that a plot of log power versus log screw speed will have a slope equal to 1 + n, so the flow index can easily be estimated.

Once again, as long as geometric similarity of the screw channel cross section is maintained between the pilot and plant scales, an experimental determination of Eq. 33 on pilot plant equipment will be predictive of any scale of extruder. So, for the case of isothermal operation, or constant extrudate properties, and for a filled screw, a method for scale-up of power has been defined.

Partially Filled Screws

Thus far it has been assumed that the extruder screw is filled with extrudate. In many cases (16,32,41,42), particularly in twin-screw extruders, this is not the case. These extruders operate starved because feed rate to the extruder is independently controlled by an external feeder. This adds to the difficulty of analysis of this situation. The basic differential equation, Eq. 24, still holds for any section of the screw that is filled with extrudate and has constant cross section geometry. The result of dimensional analysis is unchanged, but the dimensionless term containing screw length becomes the dependent variable, describing filled length, and the dimensionless term containing screw output becomes an independent variable. Stated more succinctly,

$$\frac{L_f}{L} = f\left(\frac{Q}{WVb}, \pi_1, n\right) \tag{34}$$

where L_f = the length of the screw that is filled. Similarly, the dimensionless relationship for power consumption becomes

$$\frac{P_0 b^n}{mV^{n+1}WL_f} = f\left(\frac{Q}{WVb}, \pi_1, n\right) \tag{35}$$

Nonisothermal Operation

The question of the applicability of conclusions about isothermal operation to nonisothermal extruder operation, which is the most common state of extruder behavior, is pertinent. To solve this problem, the basic differential equation must be modified to account for the effect of temperature on the flow behavior of the extrudate. The flow index n is not normally a very strong function of temperature (5,12,27,28,38), and the temperature dependency of the flow consistency, m, is generally assumed to follow an Arrhenius type equation (5,12,27,28,33,38,47,50), or an exponential relationship in temperature (44). The basic differential equation now becomes

$$\frac{dP}{dx} = m_0 \frac{\partial}{\partial z} \left[e^{(E/R)/T - 1/T_0} \left(\frac{\partial v_z}{\partial z} \right)^n \right] \tag{36}$$

or,

$$\frac{dP}{dx} = m_0 \frac{\partial}{\partial z} \left[e^{A(T - T_0)} \left(\frac{\partial v_z}{\partial z} \right)^n \right] \tag{36a}$$

where

E = an activation energy
A = an empirical constant
R = the gas constant
T = the absolute temperature, which is some function of position
T_0 = a reference temperature
m_0 = an empirical constant

Note that the constant pressure drop per unit length has been replaced with a derivative of pressure with respect to length. The
assumption of a linear pressure gradient is not valid when the properties of the fluid change with position. In this description, the
temperature can be any arbitrary function of length and vertical position. Its form depends on the source of heating and/or cooling.
For adiabatic operation of the extruder, the temperature is determined by the local rate of power dissipation. More normally, it
would be controlled by both the power dissipation and the conditions present in the heating or cooling jackets.

Equations 36 and 36a can be solved, at least in principle for a
differential length dz of the slot in Figure 1. For this differential
length, a dimensional analysis, using the techniques similar to those
used for the isothermal case, yields the following result:

$$\frac{\Delta P b^{n+1}}{m_0 L V^n} \left(\frac{\partial P^*}{\partial x^*} \right) = \frac{\partial}{\partial z^*} \left[e^{ET^*/T_0 R} \left(\frac{\partial v^*}{\partial z^*} \right)^n \right] \tag{37}$$

where

ΔP = total extruder pressure rise
$P^* = P/\Delta P$
$T^* = (T_0 - T)/T_0$

or

$$\frac{\Delta P b^{n+1}}{m_0 L V^n} \left(\frac{\partial P^*}{\partial x^*} \right) = \frac{\partial}{\partial z^*} \left[e^{AT_0 T^*} \left(\frac{\partial v^*}{\partial z^*} \right)^n \right] \tag{37a}$$

where

$$T^* = (T - T_0)/T_0$$

These equations can now be integrated with respect to length to obtain a dimensionless equation for screw output in terms of total screw back pressure.

$$\frac{Q}{WVb} = f_1\left(\frac{\Delta Pb^{n+1}}{m_0 LV^n}, \frac{ET^*}{RT_0}, n\right) \tag{38}$$

or

$$\frac{Q}{WVb} = f_2\left(\frac{\Delta Pb^{n+1}}{m_0 LV^n}, AT_0 T^*, n\right) \tag{38a}$$

It is not difficult to see from the formulation of Eqs. 38 and 38a that if corresponding points in any two extruders have the same temperatures, these equations reduce to

$$\frac{Q}{WVb} = g\left(\frac{\Delta Pb^{n+1}}{m_0 LV^n}, n\right) \tag{39}$$

Equation 39 is identical to Eq. 28. The same conclusion will be reached if the analysis is performed for a dimensionless description for total power consumption. This means that conclusions from the simple analysis, which assumed that the extruder operated isothermally, will be valid if thermal similarity is attained. The question remains as to what is required to attain this similarity.

Adiabatic Operation

When considering adiabatic operation, the temperature profile is completely defined by the power dissipation of the extruder. Analyses of this problem are found in the literature (44,51,55). The analyses will be used to draw some conclusions about the temperature profile for this situation. The rate of temperature rise during adiabatic operation is given by (44)

$$\rho C_p Q \frac{dT}{dx} = \frac{dP_0}{dx} \tag{40}$$

where C_p = the heat capacity of the extrudate and ρ = the density of the extrudate.

The gradient of power consumption with respect to length is defined by the differential form of Eq. 29.

$$\frac{dP_0}{dx} = mWV\left(\frac{\partial v_x}{\partial z}\right)^n_{z=b} \tag{41}$$

The volumetric flow rate may be replaced with the integration of the local velocity across the channel.

$$\rho C_p\left[W\int_0^b v_x \, dz\right]\frac{dT}{dx} = mVW\left(\frac{\partial v_x}{\partial z}\right)^n_{z=b} \tag{42}$$

This integral-differential equation must be solved simultaneously with Eqs. 36 or 36a. If the temperature of the feed is chosen as the reference temperature, the dimensional analysis, using Eq. 36a, yields

$$V^* = g_1\left(AT_0, \ \frac{\Delta P}{\rho C_p T_0}, \ \frac{\Delta P b^{n+1}}{m_0 LV^n}\right) \tag{43}$$

$$T^* = g_2\left(AT_0, \ \frac{\Delta P}{\rho C_p T_0}, \ \frac{\Delta P b^{n+1}}{m_0 LV^n}\right) \tag{43a}$$

Both the extrudate velocity and the temperature profile are defined by the same dimensionless groups. As a consequence, if two geometrically similar extruders are operated adiabatically at the same rotation rate and discharge pressure, thermal similarity will be attained. Under these constraints, Eq. 28, derived for the isothermal case, will apply.

External Heat Exchange

What about the more complicated situation where part of the temperature changes in the extrudate are caused by external heat sources or sinks (jackets or cored screws)? The answer to this question is not so easily formulated.

If the material within the screw is well mixed, or most of the temperature changes are attributed to viscous dissipation, then, at a fixed feed and discharge temperature, reasonably similar temperature profiles will exist on both the pilot plant and plant scales. Although single-screw extruders do a poor job of mixing the material within the screw, for many situations, most of the thermal effects are the result of the viscous heating of the material. Many twin-screw extruders, particularly those of the corotating type, do an excellent job of mixing the material within the screw. Based on these observations, the conclusion is that the dimensionless equations presented thus far are sufficient, certainly as a first approximation.

To design heating and cooling systems to control the temperature profile in the extruder, scaling-up the convective heat transfer within the extruder is required. Again referring to the simplified parallel plate model of the extruder, the differential equation describing heat transfer may be written.

$$V_x \rho C_p \frac{dT}{dx} = k \frac{\partial^2 T}{\partial z^2} + m \left(\frac{\partial v_x}{\partial z}\right)^{n+1} \qquad (44)$$

where k = the thermal conductivity of the extrudate.
The boundary conditions that apply are

$$T = T_0, \quad \text{at } x = 0 \qquad (44a)$$

$$T = T_b, \quad \text{at } z = b \qquad (44b)$$

$$T = f(x), \quad \text{at } z = 0 \qquad (44c)$$

The boundary condition at the surface of the screw is not well defined. The literature (39,44,45) suggests that the boundary condition lies between the limits of the assumption that the screw is at the initial extrudate temperature and the assumption that no heat is conducted into or from the screw. An exact determination is unnecessary. Using the same definitions as used earlier, the differential equation may be scaled and normalized to obtain

$$\left(\frac{\rho C_p V b^2}{kL}\right) z^* \frac{dT^*}{dx^*} = \frac{\partial^2 T}{\partial z^{*2}} + \frac{mV^{n+1}}{k(T_0 - T_b)b^{n-1}} \qquad (45)$$

Two dimensionless groups have appeared. The dimensionless group, $C_p V b^2 / kL$, is usually called a Graetz number. Its physical interpretation is the ratio of heat transfer by bulk flow to heat transfer by conduction. This may be put into a more familiar form:

$$Gr = Pe(b/L) \qquad (46)$$

where Pe = the Peclet number.
The dimensionless term, $mV^{n+1}/k(T - T_b)b^{n-1}$, is known as the Brinkman number. Its physical interpretation is the relative importance of heat transfer by viscous dissipation to heat transfer by conduction.

The solution to Eq. 44 can be represented by a function of the form

$$T^* = f(Gr, Br, x^*, z^*, v^*) \tag{47}$$

where Gr = the Graetz number and Br = the Brinkman number.
Equation 27 already defines dimensionless velocity in terms of other dimensionless groups, so Eq. 47 may be modified to the following form:

$$T^* = f(Gr, Br, x^*, z^*, \pi_1, n) \tag{48}$$

As for the velocity profile within the screw, the extension of Eq. 48 into a form that includes the actual geometry of the screw will result in the addition of geometric ratios.

The design engineer is interested in the average convective heat transfer coefficient. This is defined by

$$q = h(T_0 - T_b) \tag{49}$$

where h = the heat transfer coefficient and q = the heat flux.

Assuming that the extruder is at steady state, a dimensionless equation for the convective heat transfer coefficient can be readily derived from Eq. 49. At steady state, the heat transfer by convection is identically equal to the heat transfer by conduction at the surface. This leads to

$$\left(-\frac{k \, dT}{dz}\right)_{z=b} = h(T_0 - T_b) \tag{50}$$

Applying the scaling factors used earlier, the following result is obtained.

$$\frac{hb}{k} = \left(\frac{\partial T^*}{\partial z^*}\right)_{z^*=1} = Nu_x \tag{51}$$

The dimensionless group, hb/k is the Nusselt number, which is interpreted as the ratio of convective heat flow to conductive heat flow (4,23,30,34). In this situation, the natural definition of a Nusselt number appears to be a dimensionless heat flux at the wall. The same factors which affect the dimensionless temperature profile affect the Nusselt number so a dimensionless formulation for this variable should be

$$Nu_x = f(x^*, Gr, Br, \pi_1) \tag{52}$$

The Nusselt number in Eq. 52 has a local value, which is not normally the form used in design calculations. The average Nusselt number over the screw length is obtained by integration of Eq. 52 over the length of the screw.

$$Nu_{avg} = g(Br, Gr, \pi_1) \tag{53}$$

Now the task is quite straightforward. On the pilot plant scale, the extrusion conditions should be varied and the heat transfer coefficients measured. The results are then correlated in the form suggested by Eq. 53. These results are valid for a geometrically similar extruder of any size.

A point should be made about the simplifications inherent in this analysis. As before, the properties of the extrudate have been assumed not to be a function of temperature. To minimize the effects of this simplification, the experiments should be run in such a way that the extrudate temperature at the inlet and outlet of the extruder do not deviate far from the conditions actually expected on the full scale. Such a condition allows a pseudo-"average" temperature and properties of the extrudate to be assumed.

The solution to Eq. 53 has been considered in the literature (45). For the simplified case which assumes that there is no pressure flow in the extruder, the data in Figure 3 (39) have been reported. The authors know of no equivalent data that has been reported for twin-screw extruders.

E. Freezing of Foods

Consider the classical problem of the freezing of an object. Although not necessarily considered a scale-up problem, this example will illustrate the power of dimensional analysis in combination with some physical intuition.

Consider the infinite slab in Figure 4. For the purposes of simplification, we assume that external heat transfer resistance is not significant in this problem, a situation often encountered in cryogenic and air blast freezers.

Two differential equations are needed to describe this problem: one for conduction through the frozen portion and one for conduction through the unfrozen portion. The differential equations are identical, except for the application of a subscript. Assuming that the product is at its freezing point at time zero, the applicable differential equation is

$$\frac{\partial T_i}{\partial t} = \alpha_i \frac{\partial T^2_i}{dz^2} \tag{54}$$

where

    T = the temperature
    t = the time

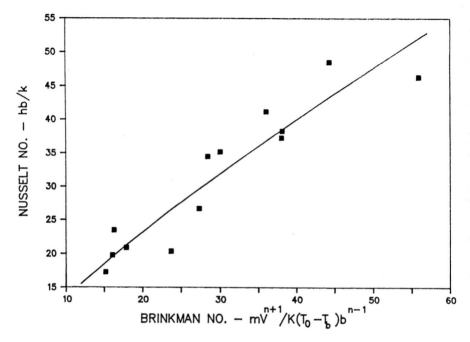

FIGURE 3   A correlation for single-screw extruder heat transfer
coefficient.   (From Ref. 39, used with permission.)

i = the water (w) or ice (f) phases
α = the thermal diffusivity
z = the direction of heat flow

The appropriate boundary conditions are:

at z = 0, T = T_∞ (54a)

at z = D, dT/dz = 0 (symmetry)   (54b)

at t = 0, T = $T_{freezing}$ (54c)

at t = 0, l = 0   (54d)

where l = the thickness of the ice layer, and D = 1/2 the slab thickness.

    In addition to these boundary conditions, an equation that couples
the two differential equations is necessary.   This equation, which

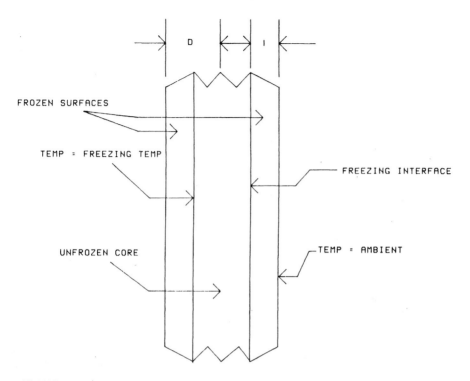

FIGURE 4   Description of freezing of a slab.

is readily obtained by a heat balance at the freezing interface, defines the growth rate of the ice layer.

$$\rho L \frac{dl}{dt} = k_w \left(\frac{dT_w}{dz}\right)_{z=1^-} - k_f\left(\frac{dT_f}{dz}\right)_{z=1^+} \tag{55}$$

where L = the latent heat, ρ = the density of ice; and $k_{w,f}$ = the phase conductivity.

To render this problem dimensionless, the procedure illustrated for simple conduction can be followed. This leads to the following scaling factors:

$$T^* = (T - T_{freeze})/(T_\infty - T_{freeze}) \tag{56a}$$

$$t^* = \alpha_w t/D^2 \tag{56b}$$

$$z^* = z/D \tag{56c}$$

$$l^* = l/D \tag{56d}$$

The dimensionless differential equations are then:
For the water,

$$\frac{\partial T_w}{\partial t^*} = \frac{\partial^2 T_w^*}{\partial z^{*2}} \tag{57}$$

For the ice,

$$\frac{\partial T_f^*}{\partial t^*} = \frac{\alpha_f}{\alpha_w} \left(\frac{\partial^2 T_f^*}{\partial z^{*2}} \right) \tag{57a}$$

And the boundary conditions are transformed to

at $z^* = 0$, $T^* = 1$ \hfill (58a)

at $z^* = 1$, $dT^*/dz^* = 0$ \hfill (58b)

at $t^* = 0$, $T^* = 0$ \hfill (58c)

$$\frac{dl^*}{dt^*} = \frac{T_f - T_\infty}{L/C_p} \left[\left(\frac{dT_w^*}{dz^*} \right)_{z^*=1^*} - \frac{k_f}{k_w} \left(\frac{dT_f^*}{dz^*} \right)_{z^*=1^*} \right] \tag{59}$$

The dimensionless group, $C_p(T_{freeze} - T_\infty)/L$, that appears in the final boundary condition is the reciprocal of the Stephan number or the enthalpy number. In addition to the Fourier number, any non-steady state heat transfer problem involving a change in state will incorporate this dimensionless group.

Inspection of the transformed equations and boundary conditions shows that the solution to the pair of differential equations will have the following form:

$$l^* = g\left(\frac{\alpha_f}{\alpha_w}, \frac{\alpha_w t}{D^2}, \frac{L}{C_p(T_f - T_\infty)}, \frac{k_f}{k_l} \right) \tag{60}$$

This does not seem to be of any immediate value, but consider some simplifications. Since the properties of water and ice are simple ratios of each other and these ratios are approximately constants, Eq. 60 can be simplified to

$$l^* = h\left(\frac{L}{C_p(T_f - T_\infty)}, \frac{\alpha_w t}{D^2} \right) \tag{61}$$

Physical intuition is useful for extracting further information from Eq. 61. Recalling that l* = 1 when the object is frozen, rearranging Eq. 61 yields

$$\frac{\alpha_w t_{total}}{D^2} = f\left(1, \frac{L}{C_p(T_f - T_\infty)}\right) \qquad (62)$$

Note that the dependence on the square of the linear size of the object once again appears. Now consider a hypothetical situation where the conductivity, density, and heat capacity of water and ice are fixed, but the heat of fusion is allowed to change. Then the time to freeze the object changes in direct proportion to the total quantity of enthalpy to be removed from it. Thus, the Stephan number, which contains the heat of fusion to the first power, must appear in the final solution as a proportionality constant. Equation 62 then becomes

$$t_{total} \; \alpha \; \left(\frac{D^2}{\alpha_w}\right)\left(\frac{L}{C_p(T_f - T_\infty)}\right) \qquad (63a)$$

Compare this to the widely used Planck equation (13,20,29,46):

$$t_{total} = \frac{\rho L}{(T_f - T_\infty)}\left[\frac{2D}{h} + \frac{D^2}{2k}\right] \qquad (63b)$$

For the case that has been assumed (no limiting external heat transfer resistance, h = ∞), the solutions agree.

For objects of other shapes, Eq. 63 is modified by adding geometric ratios of the squares of key dimensions. The literature (13,28,51) provides these correction factors. As before, if the test object is geometrically similar to the object being considered on the ultimate scale, these terms are of no significance to the designer. Consider the following example.

Suppose that experimentation reveals that freezing of small peas, having a diameter of 9 mm, in a fluid bed freezer requires 3 minutes. How long a residence time would be required to freeze sweet peas, having a diameter of 16 mm? The calculation is

$$t_{16 \; mm} = (t_{8 \; mm})(16/9)^2 = 9.5 \text{ minutes} \qquad (64)$$

IV. BUCKINGHAM π THEOREM

Unfortunately, many problems encountered in the food industry are of sufficient complexity that dimensional analysis of the differential equations cannot be utilized because the appropriate differential equations and boundary conditions cannot be accurately stated. It is for these situations that the Buckingham π theorem becomes extremely useful. Although this method does not convey as much information as dimensional analysis of the differential equations, the application of physical intuition often leads to useful solutions. At the very least, the result of a dimensional analysis through the use of the Buckingham π theorem yields a reasonable method for simplifying the data and determining a form of correlation of data that makes pilot or laboratory data predictive of full-scale equipment. It is to be preferred over correlations obtained from polynomial statistical models that are only useful for interpolation and not generally valid for extrapolation of scale-up. The theorem states

> The relationship between N dimensional variables, containing M fundamental dimensions, may be represented through the use of N-M dimensionless groups.

This seemingly simple statement was first expressed by Buckingham in 1914. Its meaning and application occupy many volumes of text (8,23,30,31,34,36,56). Most apply different techniques for the determination of the dimensionless groups and explore different definitions for the meaning of the term "fundamental dimensions." This will be avoided here with the intention of giving a practical approach to the application of the theorem.

Please note that the theorem tells us nothing of the nature of the solution, despite the tendency of practitioners to write the solution as a multiplication of dimensionless groups to arbitrary powers. This approach to writing the solution works because the actual solutions to many engineering problems take this form and this provides a great deal of flexibility in curve fitting. In addition, the theorem tells us nothing about the nature of the dimensioness groups—they need not have any physical significance, they only need to be independent of each other. This is of no great significance, since even limited physical understanding of the problem readily leads one to choose physically meaningful dimensionless groups.

The dimensional variables that the theorem refers to are the physical parameters needed to describe a problem. These usually include parameters such as viscosity, density, heat capacity, geometric size, temperature, etc.

The fundamental dimensions of the theorem are the primary units of measurement that are used to quantify the variables. There are two normal system of units that are used for this type of analysis. These are

Force, mass, length, time and temperature (FMLT°)
Mass, length, time, and temperature (MLT°)

Other texts (8,31,34,36,56) explore the value of expanding the primary dimensions to include heat, a length vector in each of the primary dimensions, and two concepts of mass (one being a quantity of matter, the other being something that force acts upon). The addition of these units to the list of primary dimensions involves significant knowledge of the physical nature of the problem. In fact, the success of their application is always illustrated by problems to which the solution is already known. This text will not consider these issues again.

The difference between the FMLT° system and MLT° systems is slight. The use of the FMLT° system requires the addition of the gravitation constant, g_c, to the list of parameters required to solve the problem. As a result the number of variables is increased by one as well as the number of fundamental dimensions. The same number of required dimensionless variables will result from the theorem. Some texts (8,56) suggest that force and mass are two different, unrelated quantities, when dealing with problems where inertial phenomena (Newton's second law) are insignificant. Again this requires a degree of physical insight that may not be available. This approach will not be considered. This text uses the MLT° system of units for all the examples. It is assumed that the reader can apply the gravitational constant where necessary. For convenience, Table 1 lists the parameters that one is likely to encounter in the MLT° system of measurement.

Some discussion is required as to the nature of temperature as a fundamental dimension. As stated earlier, to apply dimensional analysis to a problem, the variable must exhibit the property of absolute relative magnitude. The Celsius and Fahrenheit scales do not exhibit this behavior, but the equivalent absolute temperature scales, Rankine and Kelvin, do. The reader should refer to the literature for a more detailed discussion of this problem (8,31,34, 36,56). In addition to the absolute temperature scales, temperature differences have the desired property. For most heat transfer problems, the important variable is not the temperature but a temperature difference.

There is another dimension that sometimes appears in dimensional analysis: the measurement of angle. Angles can be measured as degrees, radians, etc. For the purposes of dimensional analysis angles are considered to be a fraction of a complete circle that is dimensionless (8,31). This allows the inclusion of angles in the dimensional analysis simply by addition of a dimensionless group, the angle. Similarly, parameters that are inherently dimensionless,

TABLE 1   Units of Engineering Quantities
in the MLT System

Density	M/L^3
Diffusivity	L^2/T
Force	ML/T^2
Heat capacity	$L^3/{}^\circ T^2$
Heat transfer coefficient	$M/{}^\circ T^3$
Latent heat	L^3/T^2
Mass transfer coefficient	L/T
Power	ML/T^3
Pressure	M/LT^2
Surface tension	M/T^2
Thermal conductivity	$M/L^\circ T^2$
Thermal diffusivity	L^2/T
Viscosity	M/LT

such as a coefficient of variation or a coefficient of friction, will appear in the final result of dimensional analysis as themselves.

There are some exceptions to the Buckingham π theorem that may be encountered and require some mention. Sometimes, the proper number of dimensionless groups specified by the theorem cannot be identified. There are two causes for this problem. If a parameter has been identified as a key to the problem and it really is not, then no dimensionless group can be formed using this parameter. This is readily identifiable by examining the dimensions of the parameter list. If only one variable contains one of the primary dimensions, then this variable cannot be included in the dimensional analysis and should not have been included. This situation is not an issue in food processing problems. The classical examples of movement of pendulums and gravity waves illustrate this problem and are discussed in the literature (8,31,36, 56).

Another problem that leads to inability to form the correct number of dimensionless groups is caused by an improper definition of the parameters describing the problem. The usual source of this difficulty is the fact that two of the variables in the list always appear together in the physical formulation of the problem. The most obvious example of this is the consideration of expansion coefficient and gravitational acceleration in situations dealing with natural convection. These two always appear in the solution as the multiplication of one another and so only represent one parameter, thus reducing the number of parameters required to solve the problem by one. The other obvious example is consideration of heat capacity, density, and conductivity as separate variables in simple conduction problems. The correct variable in this case may be thermal diffusivity. In practice, this is readily recognized. The last difficulty that one might encounter is the conclusion when N = M, inferring that no dimensionless variables are required. This invariably means that a parameter has been left off the list.

The question that needs to be answered is: How does one choose the correct variables and appropriate dimensionless groups? The choice of the correct variables is a result of the application of thought and experience to the problem. This infers that the practitioner has a fair degree of physical understanding of the problem. With very little practice, one can generate an appropriate list. The choice of dimensionless groups can be performed by inspection as illustrated here.

A. Important Dimensionless Groups

Table 2 is a list of the dimensionless groups normally encountered in problems of food process design and scale-up. The table is organized into several columns:

A definition of the key group
Its name(s)
Its physical significance in terms of ratio of forces
The type of problem in which it will appear

The groups are composed of variables encountered in formulations of the problem definition. Some discussion is required as to the definition of the characteristic dimension, L. For most problems the characteristic length is the most obvious dimension to use. For example, in flow-through ducts and pipes, the characteristic dimension is the radius of minimum dimension of the channel. In mixing problems, the characteristic dimension may be the vessel diameter, but sometimes is chosen as the agitator diameter. For some problems, the best choice of characteristic length may not be immediately obvious. For example in the rolling of doughs, should the characteristic length be the diameter of the roll or the gap between the rolls? It is best in these cases to choose the distance across which velocity is changing. For the case of rolling of the dough, this

TABLE 2   Useful Dimensionless Groups[a]

Group name(s) (numbers)	Definition	Type of problem	Ratio of forces
Reynolds	$\dfrac{LV\rho}{\mu}$	All flows	Inertial/viscous
Froude	$\dfrac{V}{Lg}$	Vortex flows, solid flows, settling flows, tumbling flow	Inertial/gravity
Euler	$\dfrac{P}{\rho V^2/2}$	Pressure loss, pressure development	Viscous energy/kinetic energy
Prandtl	$\dfrac{C_p\mu}{k}$	Heat transfer	Momentum differential/thermal differential
Weber	$\dfrac{V^2L\rho}{\sigma}$	Two-phase flow	Inertial/surface
Nusselt, Biot	$\dfrac{hL}{k}$	Heat transfer	Convection/conduction
Fourier			
Heat	$\alpha t/L^2$	Non-steady-state, heat transfer	
Mass	$\mathcal{D}t/L^3$	Diffusion	
Schmidt	$\mu/\rho\mathcal{D}$	Mass transfer	Momentum differential/diffusivity
Peclet	$\dfrac{\rho C_p VL}{K}$	Heat transfer	Bulk heat flow/conduction
Sherwood	$\dfrac{kL}{\mathcal{D}}$	Mass transfer	Convection/diffusion
Brinkman	$\dfrac{\mu V^2}{k\Delta T}$	Heat transfer in high viscous systems	Heat generation/conduction

(continued)

TABLE 2   (Continued)

Group name(s) (numbers)	Definition	Type of problem	Ratio of forces
Power, Lagrange group	$\dfrac{P_0}{\rho V^3 L^2}$	Power transmission	Power consumed/ inertial
Delivery, circulation	Q/VL	Continuous flow	Flow/capacity
Stephan, enthalpy	$\dfrac{\rho H}{C_p \Delta T}$	Heat transfer with state change	Latent/sensible heat

[a]The following nomenclature has been used: h = a heat transfer coefficient; H = a latent heat; k = thermal conductivity; L = a characteristic length; P = a pressure or pressure change, P_0 = a power consumption; Q = a volumetric flow rate; t = time; V = a characteristic velocity; C_p = heat capacity; ρ = density; μ = viscosity; σ = surface tension; α = thermal diffusivity;   = diffusivity; ΔT = a characteristic temperature difference.

would infer the gap between the rolls.   This is not an essential choice, but often makes interpretation of the results of the dimensional analysis easier.

The characteristic velocity is an average velocity which appears in the formulation of the problem.   For problems involving rotation the characteristic velocity used in Table 2 is customarily replaced by ND, the rotational rate times the characteristic length of the problem.   Table 3 illustrates how many of the dimensionless groups are modified for rotational systems.

Non-Newtonian behavior is exhibited by many food materials. How does one deal with this situation?   There are two approaches to this problem.   The most obvious is the replacement of the viscosity terms in the characteristic equations with an apparent viscosity.   Several examples of this will be illustrated.   An alternative approach is the redefinition of many of the dimensionless groups. For power law fluids, this is readily accomplished.   Table 4 illustrates the redefinition of many of the dimensionless groups for materials that obey the power law.   Remember that when dealing with a power law fluid, the flow index n becomes an additional dimensionless term required to describe the solution.

TABLE 3   Redefinition of Dimensionless Groups for
Rotational Problems[a]

Group name(s) (numbers)	Definition	Type of problem	Ratio of forces
Reynolds	$\dfrac{ND^2\rho}{\mu}$	All flows	Inertial/viscous
Froude	$\dfrac{N^2D}{g}$	Vortex flows, solid flows, settling flows, tumbling flow	Inertial/gravity
Euler	$\dfrac{P}{(ND)^2/2}$	Pressure loss, pressure development	Viscous energy/kinetic energy
Weber	$\dfrac{N^3D^2\rho}{\sigma}$	Two-phase flow	Inertial/surface
Peclet	$\dfrac{\rho C_p ND^2}{k}$	Heat transfer	Bulk heat flow/conduction
Brinkman	$\dfrac{\mu(ND)^2}{k\Delta T}$	Heat transfer in high viscous systems	Heat generation/conduction
Power, Lagrange group	$\dfrac{P}{\rho N^3 D^5}$	Power transmission	Power consumed/inertial
Delivery, circulation	$\dfrac{Q}{ND^3}$	Continuous flow	Flow/capacity
Blend, homochronicity	Nt	Mixing problems	Number of turnovers

[a]N = rotational rate of the system.

For other types of fluids additional dimensional groups are required to describe behavior. In general, for each additional parameter needed to describe the non-Newtonian behavior of a material, a new characteristic dimensionless group will appear in the

TABLE 4   Redefinition of Dimensionless Groups for
Power Law Fluids[a]

Group name(s)	Definition
Reynolds	$\rho L^n V^{2-n}/m$
Prandtl	$C_p m(V/L)^{n-1}/k$
Schmidt	$m(V/L)^{n-1}/\rho \mathscr{D}$
Brinkman	$mV^{1+n}/k\Delta TL^{n-1}$

[a]A power law fluid is defined by the following rheological equation: $\tau = m\dot{\delta}^n$, where τ = shear stress;
$\dot{\delta}$ = shear rate; m = flow consistency; n = flow index.

analysis. For example, when dealing with viscoelastic materials, one
or more Weisenberg or Deborah numbers will appear in the final analysis. For Bingham fluids, or other materials exhibiting a yield
stress, a Bingham number or plasticity number is needed for the
complete description of the problem. Table 5 defines these dimensionless groups.

In practice, the simplest technique for the inclusion of non-Newtonian behavior in dimensional analysis is to assume that behavior is
Newtonian. The analysis proceeds on this assumption, and the final result is modified to include the correct definition of dimensionless groups for non-Newtonian behavior and the additional dimensionless groups required.

B. Flow Through Pipes and Channels

The first example problems will rework some of the previous examples. This will permit a comparison of the power of the "blind"
use of the theorem, with dimensional analysis of the underlying differential equations.

How does one proceed? It is helpful to first draw a sketch of
the process. This helps to identify the key parameters. Consider
the problem of flow through a slot.

List the dependent variable followed by the additional variables
required to describe this problem, along with the fundamental units

TABLE 5   Dimensionless Groups for Other Non-Newtonian Flows[a]

Material type	Group name(s)	Definition
Bingham fluid	Bingham no., plasticity no.	$\tau_p/\mu V$
Viscoelastic	Weissenberg no.	$\theta_R V/L$
	Deborah no.	θ_R/t

[a]A Bingham fluid is defined by $\tau = \tau_p + \mu\dot{\delta}$, where τ_p = the yield stress; μ = the viscosity; $\dot{\delta}$ = the shear rate; τ = the shear stress. The nomenclature used in the Weissenberg and Deborah number is as follows: θ_R = the characteristic relaxation time; t = time; V = a characteristic velocity; L = a characteristic length.

needed to describe the magnitude of the variable, using Table 1 if necessary. For this problem, the variables are as follows:

Variable	Description	Units
P	Pressure drop	M/LT^2
V	Average fluid velocity	L/T
b	Slot depth	L
L	Slot length	L
w	Slot width	L
ρ	Fluid density	M/L^3
μ	Fluid viscosity	M/LT

In this case, there are seven variables and three fundamental dimensions. The Buckingham π theorem dictates that 7 - 3 = 4 dimensionless groups are required to describe the solution to the problem.

The solution can be found by inspection. Table 2 is used to identify a series of appropriate dimensionless groups. Any set of groups can be used provided that each variable appears at least once, and the groups are linearly independent of each other (all the groups in Table 2 are linearly independent of each other). The best way to proceed is to use the minimal number of groups that incorporate all the physical properties and independent and dependent variables. The balance of necessary groups will be made up of geometric ratios to the characteristic length of the problem.

The only suitable group that incorporates the dependent variable, pressure drop, is the Euler number. Since this is a flow problem, the Reynolds number must be significant, so the following partial analysis has been obtained:

$$(N_{Re}, Eu, \ldots) = 0 \tag{65}$$

Comparing the variables used in the partial solution above with the complete list of variables that describe the problem reveals that all the physical properties, the velocity, the pressure drop, and slot depth have already been incorporated into dimensionless terms. The only terms unaccounted for are the geometric variables, length and width. An obvious way to incorporate these variables into the dimensional analysis is to define two groups that are the ratios of these additional lengths to the characteristic length. The final result is

$$(N_{Re}, Eu, L/b, w/b) = 0 \tag{66}$$

or

$$P/(\rho V^2/2) = g(N_{Re}, w/b, l/b) \tag{67}$$

If the slot is very wide, the term containing the width of the slot could clearly be dropped. This yields

$$(\rho V^2/2) = g(N_{Re}, L/b) \tag{68}$$

The reader should compare this result to Eqs. 20 and 21. The application of Buckingham's theorem appears to have generated considerably less information than the analysis of the differential equations. There are two ways to simplify the problem: physical intuition or problem redefinition. Physical intuition tells us that the pressure drop through the system should be proportional to the length. This implies that the term containing the length must appear to the first power in the final solution. This allows us to rewrite Eq. 68.

$$P/(\rho V^2/2) = (L/b) \cdot g(N_{Re}) \tag{69}$$

or

$$P(b/L)/(\rho V^2/2) = g(N_{Re}) \tag{69a}$$

The equivalent result would have been obtained by redefining the problem in terms of pressure drop per unit length instead of pressure drop. This reduces the number of variables and the required number of dimensionless groups by one. The obvious choice of the dimensionless group containing pressure drop would have then been the friction factor instead of the Euler number. This would yield Eq. 69a.

Further refinement can be obtained via the regime concept discussed in the literature (34). The regime refers to whether the flow is laminar or turbulent. In general, the Euler number or friction factor depends upon the Reynolds number according to the following rules:

For laminar flow,

$$f \propto 1/N_{Re} \tag{70}$$

For turbulent flow,

$$f \approx constant \tag{71}$$

The second statement is an approximation, with the actual dependency a small power (≈ -0.2) of the Reynolds number. These "rules" are true when one is dealing with any type of shear problem, such as the power number in other agitated flow systems. The source of this statement, at least for the laminar regime, can easily be seen through the dimensional analysis of the Navier-Stokes equation, as was illustrated earlier. The physical argument for the "rule" (1) for the turbulent regime is that under turbulent conditions inertial forces are very much larger than viscous forces (the Reynolds number is large), so the effect of viscosity is minimal (≈ 0). In this light, viscosity cannot appear to significantly affect stress, so the Reynolds number must appear to the zeroth power. Similar statements can be made about the relationships between the power number and Reynolds number in laminar and turbulent flow.

The knowledge of these "rules" suggests a revision of Eq. 69a. For laminar flow, the usual case in most food processing problems,

$$P(b/l)/(\rho V^2/2) = constant/N_{Re} \tag{72}$$

or

$$Pb^2/L\mu V = constant \tag{73}$$

which are the same results as Eqs. 20 and 21. If the problem involved a non-Newtonian power law fluid, Eq. 72 is modified through the addition of a flow index as an additional dimensionless group n. The general rules apply also for the correct definition of the Reynolds number. For laminar flow

$$\frac{P(b/L)}{\rho V^2/2} \propto \frac{m}{b^n V^{2-n} \rho} \tag{74}$$

For a homologous set of problems, that is, one that considers only one fluid, and where geometric similarity is maintained, for laminar flow one can write

$$Pb^{1+n}/LV^n = \text{constant} \tag{75}$$

which is the mathematical equivalent for the general rule stated earlier.

There is great value in the analysis performed above. A complicated problem has been reduced to one that requires only the determination of a proportionality constant. A minimal number of experiments would be required to establish a set of scale-up rules for flow through a channel for a particular material.

C. Cooking a Turkey

Consider the problem of cooking the "spherical" turkey that was illustrated previously. Assuming that the underlying differential equations were unknown, a list of likely variables and the associated units are

Variable	Description	Units
t	Cooking time	T
ρ	Density	M/L^3
k	Conductivity	$ML/{}^\circ T^3$
C_p	Heat capacity	$L^3/{}^\circ T^2$
r	Turkey radius	L

There are five variables and four fundamental dimensions. The theorem requires one dimensionless group to define the problem. Table 2 suggests the Fourier number, and nothing else. This implies

$$\rho C_p t/kR^2 = \text{constant} \tag{76}$$

which is the result obtained earlier.

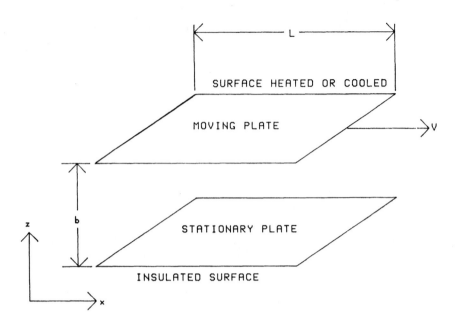

FIGURE 5   Description of heat transfer between two plates.

D. Heat Transfer Between Two Plates

Consider the problem (Fig. 5) of estimating the heat transfer coefficient for flow between two plates.   This was developed earlier as a model of heat transfer in an extruder.   A reasonable list of variables might be:

Variable	Description	Units
h	Average heat transfer coefficient	$M/°T^3$
ρ	Fluid density	M/L^3
V	Plate velocity	L/T
μ	Fluid viscosity	M/LT
d	Distance between the plates	L
L	Length of the plates	L
C_p	Fluid heat capacity	$L^3/°T^2$
k	Thermal conductivity	$ML/°T^3$

There are eight variables and four fundamental dimensions, which leads to four dimensionless groups. Referring to Table 1, the obvious dimensionless group containing the dependent variable is the Nusselt number. Since this is a flow problem associated with heat transfer, the Prandtl and Reynolds numbers must appear. The analysis yields

$$\frac{hd}{k} = f\left(\frac{dV\rho}{\mu}, \frac{C_p\mu}{k}, \frac{L}{d}\right) \tag{77}$$

or

$$Nu = f(N_{Re}, Pr, L/d) \tag{78}$$

Compare this result to Eq. 53, and note that the Brinkman number is missing. The difficulty is that there is nothing in the list of variables which incorporates the viscous dissipation term that naturally arises in the underlying differential equation. This is not a problem for most of the heat transfer problems encountered where viscous dissipation was assumed to be insignificant, but it is a major issue that cannot be ignored in this case.

There is a simple solution to the dilemma. If one knows in advance that viscous dissipation is insignificant, Eq. 77 is correct. However, what if that is not known? The only thing that can be done is to add another term to the list of variables. Because something about the physics is known, the characteristic temperature difference is added to the list of variables. This then requires an additional dimensionless group to satisfy the Buckingham π theorem. The final result is

$$Nu = g(N_{Re}, Pr, Br, L/d) \tag{79}$$

This is one of the difficulties of the method to which there are no satisfactory solutions or rules of thumb. The only suggestion that can be made is to always assume that viscous dissipation is significant and include a characteristic temperature difference in the analysis. After some data is obtained, it is relatively easy, through the use of statistical analysis of the data, or through observation of the percent of heat transfer occurring because of viscous dissipation, to determine if the inclusion of the Brinkman number is necessary.

For any simple flow situation in channels, the dimensional analysis yields the same result. The only difference between the variable list above and the list for different flow situations is that the characteristic velocity is no longer the velocity of the plate but the average velocity of the fluid being pumped through the system. As a result, identical dimensionless groups appear.

Compare Eq. 78, the solution without viscous dissipation included to Eq. 53, derived from the differential equation where flow is laminar and viscous dissipation is insignificant. From the analysis of the differential equation,

$$Nu = h(Gr) = h(N_{Re} \cdot Pr \cdot d/L) \tag{53}$$

From the application of the Buckingham π theorem,

$$Nu = f(N_{Re}, Pr, L/d) \tag{78}$$

The analysis of the differential equation has yielded much more information. Not only has it shown that the result is a function of the L/D ratio, Reynolds number, and Prandtl number, it indicates that these three variables appear as a multiple of each other, thus reducing the number of dimensionless variables which must be considered. In addition it indicates that under the laminar flow condition, viscosity does not affect the heat transfer coefficient. Equation 78, from the application of Buckingham's theorem, is not as robust in information.

Again, by referring to the regime concept, significant benefit can be obtained by using Eq. 78. In general, the dependence of the Nusselt number in forced convection problems on the Prandtl number is generally accepted to be small, $Nu \propto Pr^{1/3}$, in both laminar and turbulent flow (34). For the laminar and turbulent regimes, the following approximations can be applied to estimate the dependence of Nusselt number on Reynold's number. For laminar flow,

$$Nu \propto N_{Re}^{0.3 - 0.5} \tag{80}$$

For turbulent flow,

$$Nu \propto N_{Re}^{0.8} \tag{81}$$

For the turbulent regime, there is generally no dependence of the Nusselt number on L/D. For laminar flow, the dependence is weak, following L/D to the 1/3 power.

The "rules of thumb" discussed for heat transfer apply, by analogy, to mass transfer. Such approximations are very useful in determining whether or not scale-up problems are likely to be encountered before any experimental work is actually initiated.

The modification of Eqs. 78 and 79 in order to deal with non-Newtonian fluids represented by the power law presents no difficulty. As with flow problems, the characteristic dimensionless groups need to be modified for the power law form, and the flow index n becomes an additional dimensionless group.

It should be clear from the preceding examples that it is preferable to perform dimensional analysis via the fundamental differential equation rather than through the use of the Buckingham π theorem. However, in many practical situations there is no choice and one is forced to resort to the Buckingham π theorem.

E. Simple Mixing of Fluids

A common problem is the blending of a minor, soluble, component, such as a colorant or flavorant, into a bulk liquid. If one performs this task in a simple agitated tank as illustrated in Figure 6, the sensor detecting composition versus time will provide an output such as that shown in Figure 7. The most common method of specifying the mixing time is the time for the minor component to reach some arbitrarily defined acceptable variation (e.g., a reduction in the noise by 99% of its initial value) and then to define the mixing time as the time required to reach this limit. Alternatively, the percent variation could be considered as a dependent variable. Assume that a mixing time is specified. For this problem it can be assumed that adequate mixing is the primary process result, so

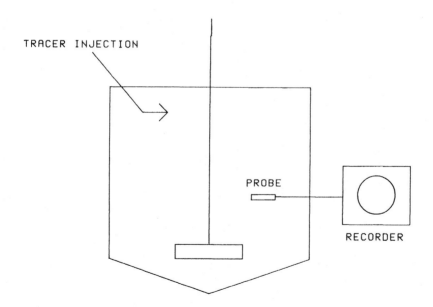

FIGURE 6 A very simple mixing experiment.

this is a case where dimensional analysis can be used to determine the primary scale-up criteria through the use of one scale of experimentation.

A reasonable list of pertinent variables might be as follows:

Variable	Description	Units
t	Mix time	T
N	Rotational speed	1/T
T	Tank diameter	L
D	Agitator diameter	L
z	Description of tank bottom	L
μ	Fluid viscosity	M/LT
ρ	Fluid density	M/L^3
θ	Agitator blade pitch	—
#	Number of blades	—
##	Number of agitators	—
L	Liquid depth	L
w	Blade width	L
g	Gravitation acceleration	L/T^2

Gravitation acceleration has been included because the tank is unbaffled and vortexing can occur. If the tank was baffled, this factor should not be included (23,48,57) but geometric factors such as the number of baffles and the width of the baffles should be included. Geometric factors such as angle of the propeller shaft or eccentricity of the propeller location could be added.

The preceding list contains 13 variables and three fundamental dimensions so ten dimensionless groups are required. This appears to be a formidable task, but is relatively straightforward. From Table 3, the obvious dimensionless group that contains the dependent variable is the Blend number (homochronicity number). The number of propeller blades, number of propellers, and blade pitch are dimensionless, so those variables are readily accounted for. This is a flow situation so the Reynolds number must be applicable. There is only one dimensionless group listed that includes gravitational acceleration, so thus far the analysis yields

$$Nt = f\left(\frac{ND^2\rho}{\mu}, \frac{N^2D}{g}, \ldots\right) \tag{82}$$

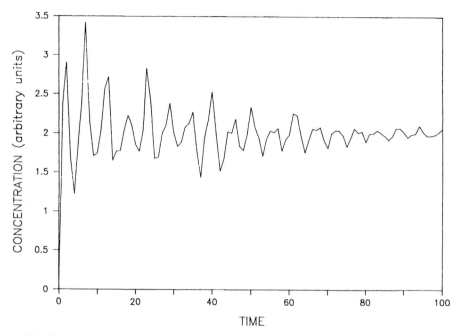

FIGURE 7 Tracer concentration versus time.

The only variables unaccounted for in Eq. 82 are geometric terms. The complete dimensional analysis must be

$$Nt = f\left(\frac{ND^2\rho}{\mu}, \frac{N^2D}{g}, \theta, \frac{L}{D}, \frac{T}{D}, \frac{w}{D}, \frac{z}{D}, \#, \#\#\right) \qquad (83)$$

Designing the full-scale process requires the execution of a simple experiment. A suitable tank geometry is chosen and several experiments are run at different agitator speeds and the mix time is measured at each condition. The results are plotted in the manner suggested by Eq. 82. This plot is completely predictive of the full-scale process.

Figures 8 (57) and 9 (44) provide correlations of the mixing time suggested by Eq 83. These figures include some of the geometric factors. The Froude number appears to be insignificant.

These correlations are valid for blending a small quantity of material in the bulk of the fluid. Figure 10 (44) illustrates that for significant quantities of the added material, the mixing time may be increased significantly.

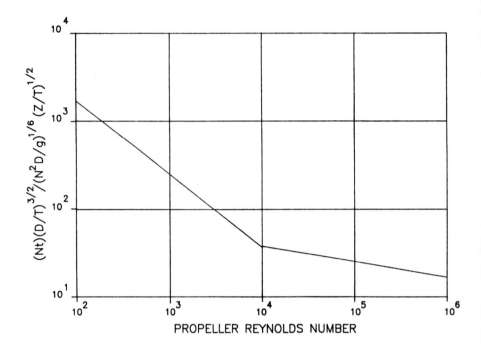

FIGURE 8   A correlation for mixing with a marine propeller.   (From Ref. 57, used with permission.)

Before proceeding to the treatment of non-Newtonian behavior, consider the secondary criteria necessary to design this process.
Power is required to turn the propeller performing the blending operation.   A correlation between mixer scale and speed is needed to specify the size of the motor required for this operation.
The variable list is identical to that given for the mixing time, but mixing time is replaced with power as the dependent variable. The analysis still requires ten dimensionless groups.   Table 2 indicates that the power number (Lagrange group) is the appropriate dimensionless group incorporating power consumption.   The result is seen to be

$$\frac{\rho}{\rho N^3 D^5} = g\left(\frac{N^2 D}{g}, \frac{ND^2 \rho}{\mu}, \theta, \frac{L}{D}, \frac{w}{D}, \frac{z}{D}, \frac{T}{D}, \#, \#\#\right) \qquad (84)$$

Figures 11 and 12 (48) illustrate this correlation for various geometries.   The Froude number is not significant in the laminar flow

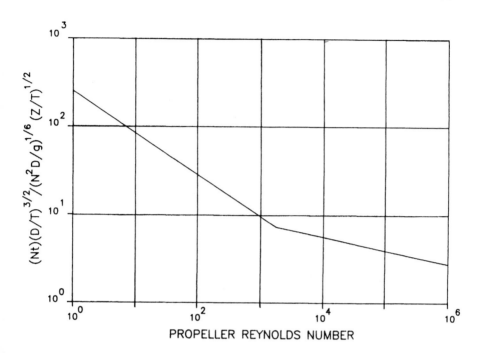

PROPELLER REYNOLDS NUMBER

FIGURE 9   A correlation for mixing with a 6-blade turbine.   (From Ref. 44, used with permission.)

regime.   The plots follow the general rule of regimes.   At low Reynolds numbers (laminar) flow the power number is inversely proportional to Reynolds number.   At high Reynolds numbers, the power number is essentially independent of Reynolds number.   Both of these conclude with the regime mentioned earlier.

Heat transfer could be another area of concern.   The parameter sought is a convective heat transfer coefficient.   The variable list becomes

Variable	Description	Units
h	Heat transfer coefficient	$M/^\circ T^3$
N	Rotational speed	$1/T$
T	Tank diameter	L
D	Agitator diameter	L

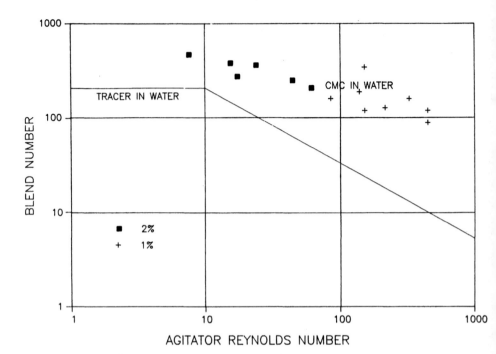

FIGURE 10 A correlation for mixing with a helical screw impeller. (From Ref. 44, used with permission.)

Variable	Description	Units
z	Clearance of tank bottom	L
μ	Fluid viscosity	M/LT
ρ	Fluid density	M/L^3
C_p	Heat capacity	L^3/°T^2
k	Conductivity	ML/°T^3
θ	Agitator blade pitch	—
#	Number of blades	—
##	Number of agitators	—
L	Liquid depth	L
w	Blade width	L
g	Gravitation acceleration	L/T^2

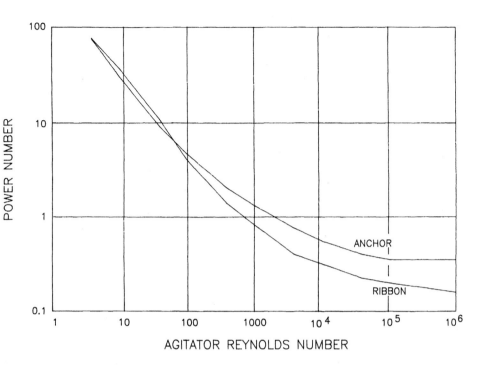

FIGURE 11 A correlation for power consumption by anchor and helical agitators. (From Ref. 48, used with permission.)

Two parameters have been added to the list, conductivity and heat capacity, but the number of fundamental dimensions has been increased by one. As a result, 11 dimensionless groups are required. The Nusselt number is the group incorporating the new dependent variable, and the Prandtl number is added to the list of independent dimensionless group. The result is

$$\frac{hd}{k} = f\left(\frac{N^2D}{g}, \frac{ND^2\rho}{\mu}, \frac{C_p\mu}{k}, \theta, \frac{L}{D}, \frac{T}{D}, \frac{w}{D}, \frac{z}{D}, \#, \#\#\right) \qquad (85)$$

Table 6 (6) provides a list of correlations following the form suggested by Eq. 85.

Dealing with non-Newtonian fluids in mixers presents some difficulty. The usual approach is to replace viscosity in the correlations with an apparent viscosity. A problem arises with the calculation of the apparent viscosity, since the shear rate within the

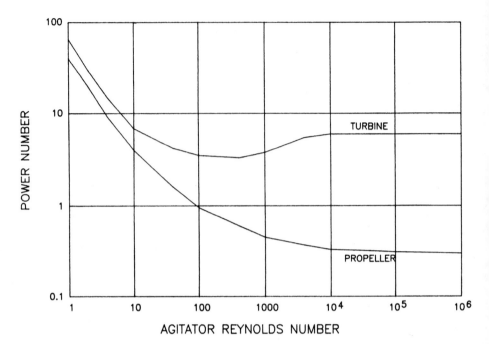

FIGURE 12   A correlation for power consumption by turbines and propeller.   (From Ref. 48, used with permission.)

mixer is an unknown function of position. Dimensional arguments (57,59) lead to the following equation, which relates an apparent shear rate to agitator speed.

$$\dot\gamma = KN \qquad\qquad (86)$$

where K = an empirical constant.

The constant in Eq. 86 is an unknown function of geometry. In general, it increases as the ratio of agitator diameter to tank diameter is increased. Table 7 (53,54,60) provides some published values for the constant in Eq. 86. This table is of limited use since many proprietary agitator designs used in the food industry are not listed.

There are two approaches to solving this problem. A generalized method suggested in the literature (59) is as follows:

Develop a dimensionless correlation of power number versus Reynolds number for the agitator of interest using a viscous

TABLE 6   Correlations for Heat Transfer in Agitated Vessels

The general form of the correlation is

$$\frac{hD_T}{k} = A\left(\frac{ND^2\rho}{\mu}\right)^a\left(\frac{C_p\mu}{k}\right)^b\left(\frac{\mu}{\mu_W}\right)^c\left(\frac{z}{D_T}\right)^d\left(\frac{D}{D_T}\right)^e\left(\frac{C}{D_T}\right)^f\left(\frac{i}{D_T}\right)^g$$

where C = clearance; D = agitator diameter; D_T = tank diameter; N = rotational rate; μ = viscosity; μ_W = wall viscosity; ρ = density; C_p = heat capacity; k = conductivity; i = agitator pitch; Z = distance off tank bottom; h = internal heat transfer coefficient.

Agitator description	A	a	b	c	d	e	f	g
Flat blade turbine								
N_{Re} < 400	0.54	0.67	0.33	0.14	—	—	—	—
N_{Re} > 400	0.85	0.67	0.33	0.14	-0.56	0.13	—	—
Retreating blade								
Baffled/6 blades	0.68	0.67	0.33	0.14	—	—	—	—
Unbaffled/3 blades								
Glassed	0.33	0.67	0.33	0.14	—	—	—	—
Unglassed	0.37	0.67	0.33	0.14	—	—	—	—
Propeller	0.54	0.67	0.25	0.14	—	—	—	—
Paddle								
N_{Re} < 4,000	0.42	0.67	0.33	0.14	—	—	—	—
N_{Re} > 4,000	0.36	0.67	0.33	0.14	—	—	—	—
Anchor								
N_{Re} < 300	1.00	0.50	0.33	0.18	—	—	—	—
N_{Re} > 300	0.36	0.67	0.33	0.18	—	—	—	—
Ribbon								
N_{Re} < 130	0.25	0.50	0.33	0.14	—	—	-0.22	-0.28
N_{Re} > 130	0.24	0.67	0.33	0.14	—	—	—	—

Newtonian fluid. (Corn syrup is a convenient material for this task.)

Obtain power numbers for several different agitator speeds using the non-Newtonian fluid of interest.

Using the correlation established for the Newtonian material, determine the Reynolds numbers that the mixer had to be operating at to obtain the power numbers measured with the non-Newtonian fluid.

Use these Reynolds numbers and the rheological model to calculate the constant for Eq. 86.

TABLE 7   Shear Rate Constants (Eq. 86) for Various Agitators

Agitator type	Impeller/tank diameter	Constant
Pseudoplastic materials		
Ribbon	0.95	30
Anchor	0.95	20–84
	0.92	72
	0.90	52
	0.84	39
	< 0.67	19
Six-blade turbine[a]	< 0.67	10–13
Two-blade paddle[a]	< 0.67	10
Curved blade paddle[a]	< 0.67	7
Three-blade propeller[a]	< 0.67	10
Dilatant materials		
Six-blade turbine[a]	0.67	50
	0.50	35
	0.33	26
Two-blade paddle[a]	0.50	27
	0.33	23
Four-blade propeller[a]	0.47	24

[a]Baffled tank

An alternative technique restricted to a single fluid may be more practical as long as experiments in the pilot plant are conducted on the same material as will be used on the full scale. The results need only be correlated by plotting the experimentally measured power number versus the non-Newtonian Reynolds number listed in Table 4. This correlation will not be predictive for other fluids, but will adequately predict the performance of any size mixer for the specific fluid. A similar approach can be used to correlate heat transfer coefficients. This technique will work for power law fluids.

Viscoelastic fluids are not treated in the literature. One would expect that a similar approach would apply. The inclusion of viscoelastic properties suggests the inclusion of a Weissenberg or Deborah number. The definition of this group for a mixer is given in

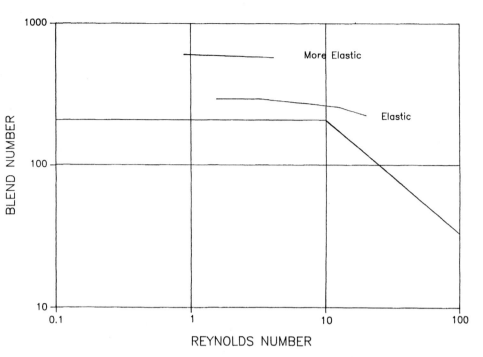

FIGURE 13   Effect of viscoelasticity on mixing with helical impellers.
(From Ref. 44, used with permission.)

Table 5.   Figure 13 (44) illustrates that for viscoelastic materials
the mixing time increases significantly. The appearance of the data
suggests that correlation of mixing times for viscoelastic materials
should follow a form such as

$$Nt = f(N_{Re})h(1 + f(N_{We}))$$ (87)

F. Mixing of Immiscible Materials

The mixing of immiscible materials is a common problem in food processing. The object of this operation is normally to increase the
interfacial area between the two phases to either reduce the tendency of the two phases to separate, such as in the preparation of
foams or emulsions, or enhance the rate of reaction beween the
two phases. The latter case was discussed in the earlier example
of gas-liquid mass transfer. How could one attack the same scale-
up through dimensional analysis?

The dependent variable of interest is the diameter of the droplet or bubble formed under the influence of agitation. A likely list of variables is prepared.

Variable	Description	Units
d_d	Droplet diameter	L
N	Rotational speed	$1/T$
T	Tank diameter	L
D	Agitator diameter	L
z	Clearance of tank bottom	L
μ_c	Fluid viscosity, continuous phase	M/LT
μ_d	Fluid viscosity, dispersed phase	M/LT
ρ_c	Fluid density, continuous phase	M/L^3
ρ_d	Fluid density, dispersed phase	M/L^3
σ	Surface tension	M/T^2
H	Holdup of dispersed phase (% of volume)	—
τ	Agitator blade pitch	—
#	Number of blades	—
##	Number of agitator	—
L	Liquid depth	L
w	Blade width	L
###	Number of baffles	—
b	Baffle width	L

Surface tension has been added as one of the variables because interfacial forces must have some effect on the size of the droplets that are formed. Additional new variables have been added to account for the properties of the dispersed phase. Gravitational acceleration has been dropped and replaced with the geometry of baffles. If this were a closed vessel (such as a colloid mill) where no vortexing could occur, gravitational forces could be neglected. The problem now incorporates 18 variables and three fundamental dimensions; thus, the analysis calls for 15 dimensionless groups.

A reasonable dimensionless group incorporating the dependent variable is the ratio of droplet diameter to tank diameter. Using

Table 2, a Reynolds number and a Weber number can be defined as pertinent groups. The dimensionless variables in the preceding list can be incorporated directly. Thus far the analysis has yielded

$$\frac{dd}{D} = f\left(\frac{N^2 D}{g}, \frac{N^2 D^3 \rho}{\sigma} \ldots\right) \tag{88}$$

Comparing the variable in the preceding equation with the variable list, it is noted that in addition to geometric variables the properties of the dispersed phase have not been incorporated. This can be accomplished in two ways: Reynolds and Weber numbers can be defined using the dispersed phase's physical properties, or the properties of the dispersed phase can be incorporated as ratios to the properties of the continuous phase. Either procedure is satisfactory, but using the ratios of the two phases' properties yields a simpler final form. The balance of the analysis is composed of geometric ratios. The final result of the dimensional analysis is

$$\frac{dd}{D} = f\left(\frac{ND^2 \rho_c}{\mu_c}, \frac{N^2 D^3 \rho_c}{\sigma}, \frac{\mu_d}{\mu_c}, \frac{\rho_d}{\rho_c}, \frac{T}{D}, \frac{b}{D}, \frac{L}{D}, \frac{z}{D}, H, \theta, \#, \#\#, \#\#\#\right) \tag{89}$$

This can be simplified by assuming homologous (34) series during scale-up. That is to say that scale-up is restricted to a geometric similar series, only one pair of fluids is being considered, and the volumetric hold-up of the dispersed phase is maintained. Equation 89 becomes

$$\frac{dd}{D} = f\left(\frac{D^2 N \rho}{\mu_c}, \frac{N^2 D^3 \rho_c}{\sigma}\right) \tag{90}$$

Some physical reasoning may be applied to this result to further clarify the problem. If the flow regime is turbulent, which it generally is in this type of problem, inertial forces predominate over viscous forces. This infers that the primary mechanism of droplet breakup is a result of turbulence (inertial stresses). Certainly at high enough Reynolds numbers, the viscous forces can be neglected, so viscosity should not appear in the final correlation. This leads to

$$\frac{dd}{D} = g\left(\frac{N^2 D^3 \rho_c}{\sigma}\right) \tag{91}$$

If a suitable technique existed for the measurement of droplet size, one scale of experimentation, under the restriction of the consideration of a homologous system, would be sufficient to establish the scale-up procedure for this process.

Note that if the assumption that viscous forces can be neglected is not made, the experimentation must find a correlation suggested by Eq. 90. In this case, the dimensional analysis requires that the experiments be run on more than one scale, or with two different fluids, because if this is not done the dependency of droplet size on the Reynolds and Weber numbers cannot be isolated.

In fact, this problem has been analyzed (57) by the fundamental consideration of the forces acting on the droplet. The conclusion that has been reached, both theoretically and experimentally, is

$$\frac{dd}{D} \propto \left(\frac{N^3 D^3 \rho_c}{\sigma}\right)^{-3/5} \tag{92}$$

Can this result be reconciled with the earlier suggestion and generally accepted view that power per unit volume should be maintained on scale-up? As illustrated earlier, power dissipation of an agitator operating in the turbulent regime is given by

$$N_{P_0} = \frac{P}{\rho_c N^3 D^5} = \text{constant} \tag{93}$$

or

$$P \propto N^3 D^5 \tag{94}$$

Substituting this result into equation 92 yields

$$dd \propto \frac{\sigma^{0.6}}{\rho_c^{0.2} (P/D^3)^{0.4}} \tag{95}$$

This demonstrates that dimensional scale-up at constant power per unit volume is identical to dimensionless scale-up at constant Weber number.

If this process involves the bubbling of a gas through an open batch of fluid, the hold-up of gas in the process is not directly known. It is not difficult to show that the holdup of gas in the batch reactor should be correlated with

$$H = \left(\frac{ND^2 \rho_c}{\mu_c}, \frac{D^3 N^2 \rho_c}{\sigma}, \frac{Q}{ND^3}, \cdots\right) \tag{96}$$

or

$$H = \left(\frac{ND^2 \rho_c}{\mu_c}, \frac{D^3 N^2 \rho_c}{\sigma}, \frac{V}{ND} \right) \tag{97}$$

In Eq. 96, the variable describing gas flow to the vessel was described by volumetric flow rate. In Eq. 97, the volumetric flow-rate was specified by superficial gas velocity (note that these are identical since $v \propto Q/D^2$).

Since the total interfacial area per unit volume is the product of hold-up (number of bubbles) and specific surface area of the bubble (bubble diameter), it is not surprising that empirical work on mass transfer in these situations identified both superficial gas velocity and power per unit volume as the key factors in scale-up. Via reasoning similar to that used to relate Weber number and power per unit volume, the literature (2,48,57) actually correlates hold-up with power per unit volume and superficial gas velocity. Figure 14 (2) presents one published result.

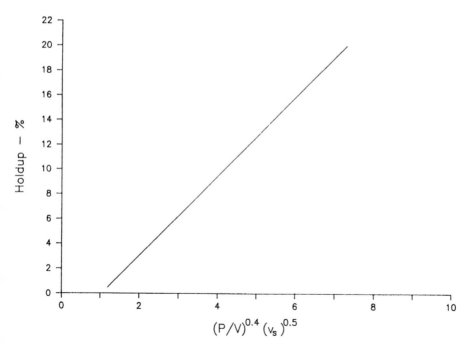

FIGURE 14   A correlation for gas hold-up.   (From Ref. 2, used with permission.)

Similar arguments could be applied to the estimation of power consumption of the agitator under gassed conditions. In general, the literature (2,48,57) does not take this approach. Instead, the power consumption of the gassed reactor is correlated with the power of the ungassed reactor and the gas flow to the reactor according to

$$\frac{P_g}{P_0} = f\left(\frac{Q}{ND^3}\right) \tag{98}$$

A plot (57) illustrating one form of this correlation is found in Figure 15.

G. Scraped Surface Heat Exchangers

Scraped surface heat exchangers are important for the processing of highly viscous food products. Two very common applications

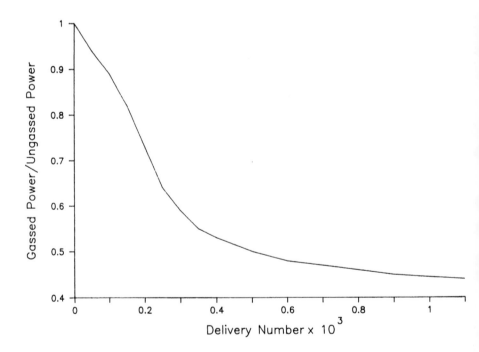

FIGURE 15 A correlation for gassed power consumption. (From Ref. 57, used with permission.)

are the freezing of ice cream and the plasticizing of shortening. Figure 16 schematically illustrates the construction of a scraped surface heat exchanger. Note that there are two primary shear fields exerted on the fluid that effect the heat transfer efficiency of this device: the rotational field induced by the motion of the shaft and blades, and the shear field caused by the longitudinal motion of the fluid through the heat exchanger. A dimensional analysis to define the convective heat transfer, assuming the fluid is Newtonian and that viscous dissipation is insignificant, follows. List the important variables:

Variable	Description	Units
h	Heat transfer coefficient	$M/{}^{\circ}T^3$
N	Rotational speed	$1/T$
D	Heat exchanger diameter	L
d	Shaft diameter	L
L	Heat exchanger length	L
μ	Viscosity	M/LT
ρ	Density	M/L^3
k	Conductivity	$ML/{}^{\circ}T^3$
C_p	Heat capacity	$L^3/{}^{\circ}T^2$
V	Liquid velocity	L/T
w	Blade width	L
#	Number of blades	—

There are 12 variables, containing four fundamental dimensions, so the correlation requires eight dimensionless groups. As before, the following equation can be written by inspection:

$$\frac{h(D - d)}{k} = f\left(\frac{ND(D - d)\rho}{\mu}, \frac{V(D - d)\rho}{\mu}, \frac{C_p\mu}{k}, \frac{L}{D}, \frac{d}{D}, \frac{w}{D}, \#\right) \quad (99)$$

Note that the difference between the shaft diameter and the shell diameter has been chosen as the characteristic length of the system because the flow occurs in a region defined by this difference. (Actually the space is half this difference, but this only has the effect of introducing a proportionality constant in Eq. 99.)

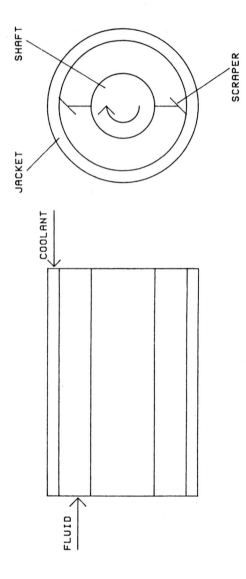

FIGURE 16 Schematic representation of a scraped surface heat exchanger.

Table 8 (15) lists some of the published correlations that follow the general form of Eq. 97. The correlations add a correction for the difference between the wall and bulk viscosity, but this is of

TABLE 8 Correlations for Scraped Surface Heat Exchangers

The form of the correlation is

$$\frac{h(D - D_S)}{k} = A\left(\frac{V(D - D_S)\rho}{\mu}\right)^a \left(\frac{ND(D - D_S)\rho}{\mu}\right)^b \left(\frac{C_p\mu}{k}\right)^c \left(\frac{\mu}{\mu_w}\right)^d$$

where h = the heat transfer coefficient; k = the conductivity; N = the rotational rate; V = average fluid velocity; ρ = density; μ = viscosity; μ_w = wall viscosity; D = heat exchanger diameter; D_S = shaft diameter.

Source	A	a	b	c	d
Bott and Romero (1963)	0.013	0.46	0.60	0.87	—
Ghosal et al. (1967)	0.123	0.79	0.65	0.60	—
Penney and Bell (1969)	0.123	0	0.78	0.33	0.018
Ramdas et al. (1977)	57.0	0.06	0.113	0.063	−0.018
Skelland (1958)	3.26	0.57	0.17	0.47	—
Skelland et al. (1962)	0.03	1.00	0.06	0.70	—
Sykora and Navratil (1966)	0.57	−0.01	0.48	0.40	—
Sykora et al. (1968)	4.09	0	0.48	0.24	—
Uhl (1966)	0.036	0	0.66	0.33	0.018
Cuevas (1982)					
Water (N_{Re} < 1,800)	0.30	0.50	0.32	0.33	0.018
Water (N_{Re} > 1,800)	0.004	0.94	0.64	0.33	0.018
Soy (N_{Re} < 1,800)	0.098	0.46	0.40	0.33	0.018

minor importance. What is important is that the equations in Table 8 omit several key factors found in Eq. 97. The terms that have been omitted are those containing the heat exchanger length, a description of the blades, and the ratio of shaft to heat exchanger diameter. These omissions may or may not be valid. Later in this section these omissions will be explored further.

There are two additional problems with the equations presented in Table 8. Most obvious is the fact that none of the equations agrees with the others. This presents the engineer with a significant difficulty: Which correlation should be used? How did this situation occur?

Some of the correlations are probably for laminar flow while others are for turbulent flow.

None of the correlations account for viscous dissipation, either in the correlation itself or, as far is known, the calculation of the heat transfer coefficients that were measured.

Geometry used to establish the correlations is not specified.

The basic form of the correlations (a power model) may not be fundamentally correct. If this is true, the correlations are an approximation of the actual functional relationship over the range of experimentation.

Assuming that the problems of inconsistency between the correlations did not exist, the problem facing the designer before applying the equations is how to modify them for non-Newtonian behavior of the fluid, since this type of heat exchanger invariably will be used for processing non-Newtonian fluids.

The last question is the easiest to answer. As with mixing, the viscosity used in Eq. 97 must be replaced with an apparent viscosity. The procedure for estimating apparent viscosity is analogous to the technique used for apparent viscosity associated with mixing problems.

1. Generate data for pressure drop versus flow using a viscous Newtonian fluid.
2. Using the data obtained in the first step, generate a correlation of friction factor versus Reynolds number for this device, such as is illustrated in Figure 17.
3. During each experimental run with actual fluid, measure the pressure drop through the heat exchanger. Calculate the friction factor associated with this pressure measurement.
4. Use Figure 17 and the calculated friction factor to derive an apparent Reynolds number. Use the definition of the Reynolds number to derive an apparent viscosity.

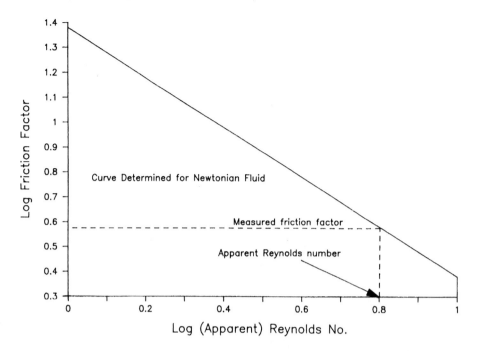

FIGURE 17 Determination of apparent viscosity from heat exchanger pressure drop.

5. Correlate the apparent viscosity just determined with fluid velocity and shaft rpm. For a pseudoplastic fluid, the result will have the appearance of Figure 18.

There is a constraint on the correlation developed in Figure 18. It only holds for a specific heat exchanger geometry, that is, a specific ratio of heat exchanger diameter to shaft diameter. The result could be generalized by measuring the apparent viscosity for various geometries. It can be shown that the correlation would then take the following form:

$$\mu_{app} = f\left(\frac{ND}{(D - d)}, \frac{V}{D - d}\right) \tag{100}$$

Determination of heat transfer coefficients for non-Newtonian fluids requires experimental data for the specific fluid. Published correlations are not adequate for this purpose.

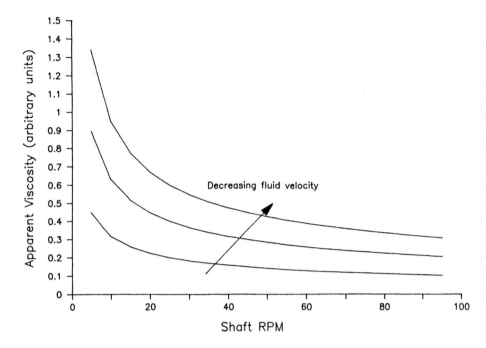

FIGURE 18   Qualitative relationship between apparent viscosity, flow rate, and rpm.

Almost any measurement of the convective heat transfer coefficient should be corrected for viscous dissipation. This is necessary because the assumption of the insignificance in heat transfer problems is generally not valid for the highly viscous materials that are normally processed in scraped surfce heat exchangers. If this is not included, the heat transfer coefficients predicted can be highly inaccurate. A complete heat balance around the scraped surface heat exchanger is:

$$Q_p + Q_c + Q_v = 0 \qquad (101)$$

where

Q_p = the total sensible and latent heat change of the product stream

Q_c = the heat transfer through the cooling or heating medium

Q_v = the heat introduced to the system via viscous dissipation

If viscous dissipation is neglected in the calculation of the heat transfer coefficient, the effect will be "lumped" in the term which accounts for the heat transfer through the heating or cooling medium. As a result the convective heat transfer coefficient will be in error. This error will not be consistent. If the material is being cooled, the removal of heat by the cooling medium will appear to be less than it actually is. Conversely, if the material is being heated, the error will be in the opposite direction. As a result, any correlation of heat transfer coefficients, in which the experimenter has not included the viscous dissipation effect, will be of questionable value. This is best illustrated by example.

Consider a scraped surface heat exchanger operating under the following conditions:

Feed rate	100 kg/hr
Heat capacity of feed	2,000 J/Kg °C
Feed temperature	100°C
Discharge temperature	20°C
Coolant temperature	0°C (assumed constant)
Net power input[2]	1 kw
Heat transfer area	1 m^2

First calculate the apparent heat transferred to the feed material. This is the sensible heat change.

$$Q_p = 2,000(100)(100 - 20) = 8,000 \text{ kJ/hr}$$
$$= 2.22 \text{ kw}$$

The log mean temperature difference is

$$T_{LM} = \frac{(T_{in} - T_{coolant}) - (T_{out} - T_{coolant})}{\ln(T_{in} - T_{coolant}) - \ln(T_{out} - T_{coolant})} = 49.7°C$$

The apparent average heat transfer coefficient is

$$U = 2.22/1/49.7 = 0.045 \text{ kw/m}^2 \text{ °C}$$

But the heat actually transferred from fluid to coolant is greater than that calculated above. The actual heat removed by the coolant is the total of the sensible heat change plus the net power

---

[2] This is the power consumption of the motor, less frictional, windage, resistance, etc., losses.

input (2.22 kw + 1 kw = 3.22 kw). The actual heat transfer coefficient is

$$U = 3.22/(1 \cdot 49.7) = 0.065 \ kw/m^2 \ °C$$

This indicates that the original estimate of heat transfer coefficient was in error by 44%.

Experimentally the viscous dissipation can be measured by direct measurement of the power being transferred by shaft work, or by a careful heat balance on both the product and heating/cooling media streams.

Considerable insight can be gained from an analysis of the underlying differential equations. This will explain some of the difficulties encountered when using the Buckingham π theorem. If one neglects the region in the immediate vicinity of the blades, this is a problem in helical flow, as is illustrated in Figure 19. The literature (5) provides an analysis of this flow situation. To simplify the problem further, assume that the gap between the shaft and the wall is small compared to the diameter of the heat exchanger. In a manner similar to that used in the analysis of extruders, the actual flow model can be approximated as flow between parallel plates, as illustrated in Figure 20. The pertinent differential equations are

$$0 = \frac{d\tau_{xy}}{dx} \tag{102}$$

$$\frac{\Delta P}{L} = \frac{d\tau_{xz}}{dx} \tag{103}$$

Subject to the boundary conditions:

$$\text{at } x = 0, \ v_y = 0 \tag{102a}$$

$$\text{at } x = 0, \ v_z = 0 \tag{102b}$$

$$\text{at } x = b, \ v_y = V \tag{103a}$$

$$\text{at } x = b, \ v_z = 0 \tag{103b}$$

The shear stresses for a power law fluid are given by

$$\tau_{xj} = -m\dot{\gamma}^{n-1} \frac{dv_j}{dx} \tag{104}$$

where the magnitude of the shear rate is given by

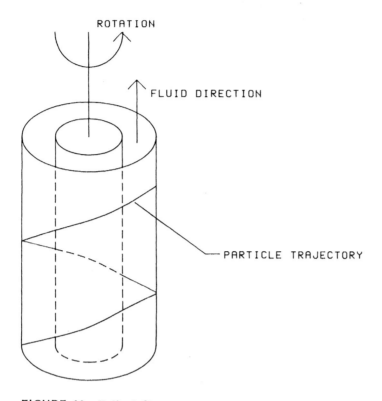

FIGURE 19 Helical flow.

$$\dot{\gamma} = \sqrt{\left(\frac{dv_y}{dx}\right)^2 + \left(\frac{dv_z}{dx}\right)^2} \tag{105}$$

Defining the following dimensionless variables:

$$\tilde{x} = x/b \tag{106}$$

$$\tilde{v}_i = v_i/V \tag{107}$$

$$\pi = \frac{\Delta P b^{n+1}}{m L V^n} \tag{108}$$

Substitute these definitions into the differential equations to obtain

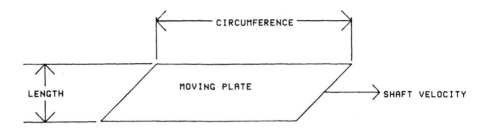

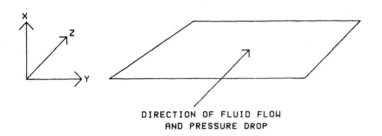

FIGURE 20   Two-plate approximation for helical flow.

$$\frac{d}{d\tilde{x}}\left(\left[\left(\frac{d\tilde{v}_y}{d\tilde{x}}\right)^2 + \left(\frac{d\tilde{v}_z}{d\tilde{x}}\right)^2\right]^{\frac{n-1}{2}}\frac{dv_y}{d\tilde{x}}\right) = 0 \tag{109}$$

$$\frac{d}{d\tilde{x}}\left(\left[\left(\frac{d\tilde{v}_y}{d\tilde{x}}\right)^2 + \left(\frac{d\tilde{v}_z}{d\tilde{x}}\right)^2\right]^{\frac{n-1}{2}}\frac{d\tilde{v}_z}{d\tilde{x}}\right) = \pi \tag{110}$$

From this one dimensionless form of the solution for the local velocity is given by

$$\tilde{v}_i = f_i(\pi, n, \tilde{x}) \tag{111}$$

This result is in terms of the pressure drop through the system, which, from a more typical point of view, is a dependent variable. The pressure term can be eliminated by resorting to the equation of continuity:

$$(v_z)_{avg}b = \int_0^b v_z\, dx \tag{112}$$

Performing the indicated integration and substituting into Eq. 111 yields a result only in terms of the independent variables.

$$\tilde{v}_i = g_i\left(\frac{(v_z)_{avg}}{v}, n, \tilde{x}\right) \tag{113}$$

Note that the variable appearing in Eq. 113 is the ratio of the rotation velocity and the velocity in the flow direction.

Consider a region close to the wall (x = 0). The fluid flow in this region predominates in the determination of the heat transfer coefficient. The velocity of the fluid, *following the helical streamlines*, is given by

$$v_p = \sqrt{V_z^2 + V_y^2} \tag{114}$$

To simplify the analysis, without any loss of important information, linearize this to

$$v_p \approx \left(\frac{dv_z}{dx}\right)_{x=0} dx + \left(\frac{dv_y}{dx}\right)_{x=0} dx \tag{115}$$

One can write a differential energy balance in a new coordinate system which approximates the actual situation. One direction is the helical path following the flow, the second direction is normal to that path, which represents the radial direction of the heat exchanger. This is also equivalent to the x direction in the parallel plate approximation. If viscous dissipation is not neglected, the following equation is obtained:

$$v_p \rho C_p \frac{dT}{dP} = k\frac{\partial^2 T}{\partial x^2} + m\left(\frac{dV_p}{dx}\right)^{n+1} \tag{116}$$

The boundary conditions for this problem are:

at P = 0, T = T_0 (116a)

at x = 0, T = T_b (116b)

at x = 1, T = f(P)   (116c)

Define the following dimensionless terms:

P* = P/L

T* = $(T - T_0)/(T_b - T_0)$

$\tilde{x}$ = x/b

Substituting the dimensionless terms into Eq. 116 yields

$$\frac{v_p \rho C_p b^2}{L} \frac{dT^*}{dP^*} = \frac{\partial^2 T^*}{\partial \tilde{x}^2} + Br \tag{117}$$

where Br = the Brinkman number. From which the dimensionless temperature is defined as follows:

$$T^* = f\left(\frac{v_p \rho C_p b^2}{L}, \ Br, \ P^*\right) \tag{118}$$

Using the same techniques as illustrated in the analysis of heat transfer in extruders, it is not difficult to show that the dimensionless equation for heat transfer coefficient is

$$Nu = g\left(\frac{\rho b^n V^{2-n}}{m}, \ \frac{mV^{1+n}}{K(T_0-T_b)b^{n-1}}, \ \frac{C_p m V^{n-1}}{b^{n-1}}, \ \frac{L}{b}, \ \frac{(v_z)_{avg,n}}{V}\right) \tag{119}$$

Note that one Reynolds number for rotational flow and a term containing the ratio between rotational and linear flow appears. Simple substitution allows the inclusion of a Reynolds number for the flow direction.

From Eq. 119, it is clear that the two Reynolds numbers are not likely to appear in the final result as the multiplication of each other to simple powers, as is assumed in the literature. In addition, the Brinkman number has appeared, although it does not appear in any of the published correlations. Look at the extremes. If the rotational velocity and viscous dissipation are negligible, the following result is obtained.

$$Nu = h_1\left(\frac{\rho b^n (V_z)_{avg}^{2-n}}{m}, \ \frac{C_p m (V_z)_{avg}^{n-1}}{b^{n-1}}, \ n, \ \frac{L}{b}\right) \tag{120}$$

If the flow velocity is small and viscous dissipation is negligible, then

$$Nu = h_2\left(\frac{\rho b^n V^{2-n}}{m}, \ \frac{C_p m V^{n-1}}{b^{n-1}}, \ n, \ \frac{L}{b}\right) \tag{121}$$

Examination of Eqs. 120 and 121 reveals that correlation of the Nusselt number with the product of the two Reynolds numbers to exponents will be satisfied by varying values for the exponents depending on the relative importance of each flow. This is probably one reason for the disparity in the values of the published constants.

To conclude, the power dissipation of the scraped surface heat exchanger should be considered, since to accomplish a complete scale-up, both heat transfer coefficient and power dissipation must be predictable. For Newtonian flow problems the solution can be written directly as

$$\frac{P}{\rho N^3 D^4 L} = f\left(\frac{ND(D - d)}{\mu}, \frac{d}{D}\right) \tag{122}$$

The literature (1) reports that at low Reynolds numbers the power number is inversely proportional to the Reynolds number and at high Reynolds numbers, the power number is constant. This is consistent with the regime concept. The result equivalent to Eq. 122 for non-Newtonian fluids is left as an exercise for the reader.

H. Mixing and Sheeting of Doughs

The mixing and sheeting of doughs are two common unit operations in the food industry, finding widespread application in the bakery, snack food, and breakfast cereal industries. Dimensional analysis of these operations reveals how potential scale-up problems can be diagnosed before any experimentation is initiated.

Consider the schematic representation of a dough mixer that is shown in Figure 21. Cereal chemists generally consider the specific energy input as being indicative of the degree of "development" of the dough. Specific energy is defined as the total energy dissipation per pound of dough. Assuming that the dough is Newtonian, which is clearly a crude approximation, perform a dimensional analysis for the power consumed by a dough mixer.

The key variables are as follows:

Variable	Description	Units
p	Power	ML/T^3
ρ	Density	M/L^3
μ	Viscosity	M/LT
N	Rotational speed	$1/T$
F	Volumetric charge	L^3
d	Roller diameter	L
b	Breaker bar diameter	L
D	Mixer diameter	L
h	Clearance between rollers and wall	L

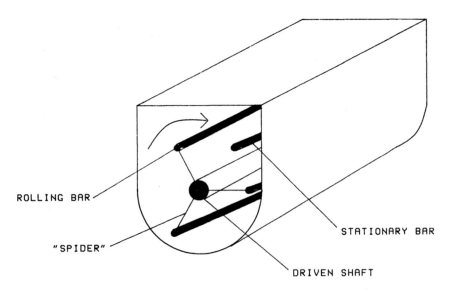

ROLLING BAR

"SPIDER"

STATIONARY BAR

DRIVEN SHAFT

FIGURE 21 Schematic representation of a dough mixer.

Variable	Description	Units
t	Clearance between breaker and rollers	L
L	Length of mixer	L

By inspection the dimensionless relationship that is sought is

$$\frac{P}{\rho N^3 D^4 L} = f\left(\frac{ND^2\rho}{\mu}, \frac{d}{D}, \frac{F}{D^3}, \frac{h}{D}, \frac{t}{D}\right) \tag{123}$$

Solving for power and making use of the general rule that in laminar flow the power is proportional to the velocity squared (see the sections about flow in pipes and channels and the behavior of extruders), it can be shown that

$$P \propto \mu N^2 D^2 L \; g\left(\frac{d}{D}, \frac{F}{D^3}, \frac{h}{D}, \frac{t}{D}\right) \tag{124}$$

The specific energy is simply the power dissipation multiplied by the mixing time divided by the mass of dough in the mixer. Then the required time of mixing is given by

$$t \propto \frac{\dot{W}}{P} \qquad (125)$$

where $\dot{W}$ = specific energy input. If the scaled-up mixer is geometrically similar to the pilot mixer, this simplifies to

$$t \propto \frac{1}{N^2} \qquad (126)$$

A conclusion can be drawn from Eq. 126. If the plant mixer operates at the same speed as the pilot plant mixer, the times required for developing the dough is identical. This provides the designer with a clear-cut direction as to how to design the scaled-up mixer. Unfortunately, for most actual applications geometric similarity is not maintained when scaling up bakery mixers.

How does one handle a real dough with non-Newtonian properties? Complete analysis is impossible since the dough exhibits very complicated, time-dependent behavior. However, one can draw some useful inferences from the non-Newtonian analyses that follow.

If the dough were a simple power law fluid, the viscosity in the list of variables would be replaced by the flow consistency and the flow index. The result of the dimensional analysis would be

$$\frac{P}{\rho N^3 D^4 L} = f\left(\frac{N^{2-n}D^2\rho}{m}, \ldots, n\right) \qquad (127)$$

The time required to mix the dough is then

$$t \propto \frac{1}{N^{1+n}} \qquad (128)$$

As before, if geometric similarity is maintained, the time required to mix the dough is held constant on scale-up if the speed of the mixer is held constant.

For a viscoelastic dough, which can be approximated as a Maxwell body, the variable list would have the relaxation time added. The new dimensionless group which appears to incorporate this new term is the Weissenberg (Deborah) number. The result is

$$\frac{P}{\rho N^3 D^4 L} = h\left(\frac{ND^2\rho}{\mu}, \theta_R N\right) \qquad (129)$$

where θ_R = the relaxation time of the dough. The conclusions are unchanged. No matter how complicated a model is considered, even a model containing an infinite number of relaxation times, the conclusion is the same.

Analysis of the dough sheeter represented in Figure 22 proceeds in much the same manner.

For a Newtonian dough the power dissipation of the roll is given by

$$\frac{P}{\rho N^3 D^4 W} = f\left(\frac{ND^2\rho}{\mu}, \frac{gap}{D}, \frac{feed}{D}\right) \tag{130}$$

Geometric similarity has no real meaning for this problem, since both the input and output thickness of the dough are normally specified. Nonetheless some conclusions about the scale-up of sheeting rolls can be drawn through the use of physical intuition.

In order to draw some inferences from the behavior of the sheeter, first make an estimate of the volumetric output of the sheeter, such as

$$q \approx NDw(gap) \tag{131}$$

where q = the volumetric output and w = width of roll.

Using this relationship and the general rule that the power is proportional to the square of characteristic velocity, an equation for the specific energy consumption of the sheeter can be written:

$$\dot{W} \propto N \tag{132}$$

Now consider the simplest type of production scale-up, simply increasing the speed of an existing device. Equation 132 shows that specific energy input of the sheeter must increase if the sheeter's speed is increased. We can conclude that if specific energy input is an important determinant of dough quality and the pilot plant development of the product was performed on full-scale rolls, the products produced at commercial sheeter speeds will be different from those produced in the pilot plant. There is no immediately obvious solution to this problem.

How can one assess the more common problem of going to larger diameter rolls for the production scale? This can be accomplished by considering the elements of power dissipation of the rolls. A force balance yields the following equation for the power dissipation of the rolls:

$$P = ND \int_{surface} \tau_{surface} d(surface) \tag{133}$$

As a first approximation, the shear rate between the rolls and the shear stress that develops as a consequence of that shear rate are approximately given by

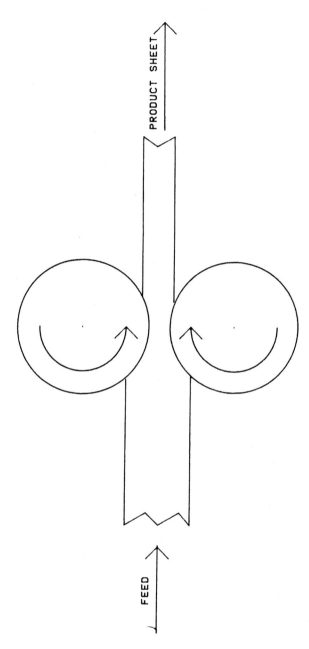

FIGURE 22 Schematic representation of dough sheeting.

$$\dot{\gamma} \approx \frac{ND}{\text{clearance}} \tag{134}$$

If the feed thickness and gap for two different diameter rolls producing the same quantity of dough are identical, it follows from Eq.
134 that the shear stress and shear rate are the same for points between the rolls that have the identical spacing. Since the velocity
of the dough on the roll surface is the same for both rolls, the variables within the integral of Eq. 133 are identical for every point
of contact with the roll, where the gaps are identical. As a result
the only variable causing the power dissipation to change is the difference in the length of contact between the rolls. What changes
the length of contact? The length of contact increases with an increase in the feed thickness and decrease in the ratio of gap to roll
diameter. If on scale-up, the gap is held constant in order to meet
a specified product thickness, larger rolls must dissipate more energy per pound of production. This may result in different plant
and pilot plant qualities. Again, there is no immediately obvious solution to this problem.

How reasonable are these conclusions? The differential equations
that describe the flow between rolls have been solved in the literature (3,7,14,17−19,21,22,24,25,35,37,43,44,54). Examination of
these solutions reveals that the qualitative conclusions that have
been drawn above are valid.

Since both increasing speed at constant diameter and increasing
diameter at constant production rate will invariably result in an increase in specific energy input, which is presumed to control product quality by the creation and/or destruction of intra/intermolecu-
lar protein bonds, how can this unit operation be satisfactorily
scaled up? Unfortunately it cannot be scaled up in the manner
normally applied to other processes.

How does one deal with this dilemma in a practical situation?
There are several approaches, the best of which is as follows:
Rather than scaling the process up, the pilot plant should be a
duplicate of the anticipated plant scale in all factors (speed, diameter, etc.), other than roll width. The roll width on the pilot
plant should be wide enough so that edge effects are insignificant,
but narrow enough so that the production rate is small enough for
the pilot process to be economically run.

How narrow the pilot plant rolls would need to be is determined
by performing an order-of-magnitude analysis on the underlying
differential equations describing the flow between the rolls.

We are not aware of any completely satisfactory approach to the
scale-up of sheeting rolls. The problem is discussed in some detail
in three papers and a recently published book (17−19,37).

Another factor the designer must consider when designing sheeting rolls is the force exerted on the rolls by the dough. This force is not insignificant and may result in a significant deflection of the rolls (19,44). As a consequence of this deflection, the final thickness of the dough will not be uniform across the width of the dough sheet. Dimensional analysis suggests the following relationships between closing force and the other variables.

$$\frac{FD^3/W}{\rho(ND)^2} = f\left(\frac{ND^2\rho}{\mu}, \frac{gap}{D}, \frac{feed}{D}\right) \tag{135}$$

where F = the closing force.

The reader is referred to any standard mechanical engineering text for instruction on how to estimate the magnitude of roll deflection that will occur because of these forces. A recent publication (19) covers this aspect of sheeting scale-up in some detail.

I. Performance of a Versator

A Versator[1] is a relatively unfamiliar unit operation that has found application in the food industry. The device (Fig. 23) is used to deaerate products. This is accomplished by feeding the product, under vacuum, to a disk that is spinning at high speed. The product spreads across the surface of the disk, and suspended air bubbles are readily removed.

There are several costs associated with this unit operation:

A significant amount of shear energy is dissipated by the rotating plate which elevates the temperature of the product. Since many components of food products may be degraded by elevated temperatures, the passage of the product through the Versator may have deleterious side effects.

The movement of the product in a thin film through a high vacuum environment will result in some distillation of volatile components of the food product. Since many flavor components are highly volatile, this could have a significant effect on the ultimate flavor of the product passed through the Versator.

The amount of viscous dissipation by the plate may be determined by measuring the power consumption of the Versator. A

---

[1]This device is manufactured by the Cornell Manufacturing Corporation, Springfield, NJ.

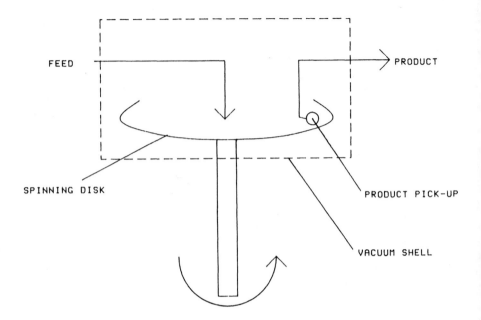

FIGURE 23   Schematic representation of a Versator.

dimensional analysis reveals that power can be related to the feed
rate to the plate, plate size, and speed by

$$\frac{P}{\rho N^3 D^5} = f\left(\frac{ND^2\rho}{\mu}, \frac{q}{ND^3}\right)$$ (136)

In this case it is not clear whether the flow regime is laminar or
turbulent, but consider either case. If the flow is laminar, then
as a general rule the power number is inversely related to the Reynolds number so that

$$\frac{P}{\rho N^3 D^5} = \frac{\mu}{ND^2\rho} g\left(\frac{q}{ND^3}\right)$$ (137)

For the turbulent case, the power number is essentially independent
of Reynold's number, so Eq. 136 becomes

$$\frac{P}{\rho N^3 D^5} = h\left(\frac{q}{ND^3}\right)$$ (138)

For the laminar case, the specific energy input, , is found to be:

$$\dot{W} = \frac{P}{q} \propto N^2 D^3 g \left(\frac{q}{ND^3} \right) \tag{139}$$

where q = the volumetric feed rate.

From this equation, it can be concluded that if the feed rate to the Versator is kept proportional to the cube of disk diameter and the rotational speed is kept constant, the specific energy input for both the pilot plant and plant disk will be identical. Hence, the temperature rise of the products will be identical.

For the turbulent case, the specific energy is found to be

$$\dot{W} = \frac{P}{q} = N^3 D^5 h \left(\frac{q}{ND^3} \right) \tag{140}$$

In this case, if the feed rate is kept proportional to the cube of the disk diameter, specific energy input will rise significantly if the speed of the disk is not reduced. Since without experimental data the exact functionality of the delivery number is unknown, the required reduction in disk speed cannot be determined. However, it is reasonable to conclude that if temperature rise of the product is a critical factor, the capacity of the Versator will be less than proportional to the cube of disk diameter.

Consider the possibility of distillation of the volatile components of the product. Defining a mass transfer coefficient as follows:

$$n_g = ka(c - c^*) \tag{141}$$

where

$\quad$ k $\quad$ = the mass transfer coefficient
$\quad$ c $\quad$ = the concentration of the distilled material
$\quad$ c^* = the equilibrium concentration
$\quad$ a $\quad$ = the area of the disk
$\quad$ n_g = molar transfer rate

A dimensional analysis yields

$$\frac{kD}{\mathcal{D}} = f \left(\frac{ND^2 \rho}{\mu}, \frac{q}{ND^3}, \frac{\mu}{\rho \mathcal{D}} \right) \tag{142}$$

where $\mathcal{D}$ = the diffusivity. Using a power law approximation for the function

$$\frac{kD}{\mathcal{D}} = A \left(\frac{ND^2 \rho}{\mu} \right)^a \left(\frac{q}{ND^3} \right)^b \left(\frac{\mu}{\rho \mathcal{D}} \right)^c \tag{143}$$

Experiments must be performed to determine the exponents in Eq. 143. If it is assumed that the delivery number Q/ND^3, is kept constant on scale-up, which seems a reasonable first approach, Eq. 143 is simplified to

$$\frac{kD}{\mathscr{D}} = B \left(\frac{ND^2\rho}{\mu}\right)^a \left(\frac{\mu}{\rho\mathscr{D}}\right)^b \tag{144}$$

If the regime concept is used, the exponent, a, in Eq. 144 must be between 0.3 and 0.8, and the exponent, b, is approximately 0.33. As a consequence a relationship for mass transfer coefficient, speed, and scale can be written.

$$k \propto \frac{1}{D} (ND^2)^a \tag{145}$$

where a ranges from 0.3 for laminar flow to 0.8 for turbulent flow.

Now, from both mechanical design and power dissipation considerations, the speed of large disks should not exceed that of small disks. Assume that speed is kept constant on scale-up, which implies that the feed rate is proportional to the cube of disk diameter.

The rate of distillation of volatile material is proportional to the product of plate area and mass transfer coefficient. It follows that the distillation of volatiles per unit throughput is given by the following: for laminar flow,

$$\dot{m} = \frac{kA}{q} \propto \frac{D^2}{D^3}\left(\frac{1}{D}\right)(D^2)^{0.3} = \frac{1}{D^{1.4}} \tag{146}$$

and for turbulent flow,

$$\dot{m} = \frac{kA}{q} \propto \frac{D^2}{D^3}\left(\frac{1}{D}\right)(D^2)^{0.8} = \frac{1}{D^{0.4}} \tag{147}$$

In either case, the fraction of volatile material removed will be less for a larger than a smaller Versator. One could conclude that if the removal of desirable components is acceptable on the small pilot-scale Versator, it will be acceptable on the large plant Versator.

J. Handling and Mixing of Solids

The scale-up of unit operations that incorporate the mixing and handling of solids is an important aspect of food processes. Some common examples in food processing that may be considered are the following:

Mixing of powders
Grinding of powders
Coating of solids
Flow of solids

Although it is not obvious, all of these operations have much in common. The flow of solids, particularly in rotational devices such as mixers and tumbling drums, is so complex as to defy fundamental analysis. As such, it is ideal for the application of dimensional analysis of the Buckingham π theorem.
Consider the mixing of two solid materials in a tumbling drum. To simplify the problem, assume that the two materials are identical in all aspects other than color. A relationship is sought for the mixing of these ingredients as a function of the variables that describe the system. First define the "quality" of mixing by a coefficient of variation which is dimensionless. The literature (57,58) is full of details about how one might measure this coefficient of variation, but that is not the issue here.
Proceed to make a list of variables which describe the system.

Variable	Description	Units
σ^2	Coefficient of variation	—
D	Drum diameter	L
d	Particle diameter	L
V	Volumetric charge of drum	L^3
g	Gravitational acceleration	L/T^2
ρ	Density of solids	M/L^3
N	Rotational speed of drum	$1/T$
t	Time of mixing	T
L	Length of mixer	L

The gravitational acceleration has been included because this process involves the rise and fall of the material. The problem incorporates nine variables and three fundamental dimensions. Buckingham's theorem states that six dimensionless groups are required to describe the process. Using the coefficient of variation as the dependent dimensionless group, the following relationship can be written

$$\sigma^2 = f\left(\frac{N^2D}{g}, \frac{V}{D^3}, \frac{L}{D}, Nt, \frac{d}{D}\right) \tag{148}$$

Equation 148 uses six dimensionless groups, but the density is not included in the relationship. Since no other term in the list of variables contains the dimension of mass, it can be concluded that density should not have been incorporated into the original list. As a result, there are actually only eight variables that use two fundamental dimensions, which still requires a relationship containing six dimensionless groups. If the volumetric fill of the drum were replaced with the mass of charge in the drum, density would have been included in the final relationship as follows:

$$\sigma^2 = f\left(\frac{N^2D}{g}, \frac{M}{\rho D^3}, \frac{L}{D}, Nt, \frac{d}{D}\right) \tag{149}$$

Equations 148 and 149 are equivalent to each other.

Several conclusions can be drawn from Eq. 148 or 149. Assume that the percent fill and L/D of the vessel is held constant upon scale-up. Furthermore, if the diameter is very large compared to the diameter of the particles being mixed, it is reasonable to assume that the particles only "see" one another and not the wall of the vessel. As a consequence the term incorporating the ratio of particle diameter to vessel diameter is not important. Under these conditions Eq. 148 reduces to

$$\sigma^2 = f\left(\frac{N^2D}{g}, Nt\right) \tag{150}$$

Now, suppose that the process requirements specify that the two ingredients are to be mixed to some arbitrary degree. The time for this mixing to occur is defined as t_{mix}. Then the functional relationship needed to scale up the process is

$$Nt_{mix} = f\left(\frac{N^2D}{g}\right) \tag{151}$$

Scale-up of this process is readily accomplished by experimentation on a single pilot plant scale. One simply needs to experimentally establish the functional relationship suggested by Eq. 151.

At first glance, it would seem that the mixing time could be made as arbitrarily short as desired by simply increasing the speed of the mixer. Intuition tells us that this cannot be true. There is a good physical explanation why this cannot be true. The Froude number in Eq. 151 is simply the number of "g's" exerted on the particles. As the Froude number increases, there is a greater and

greater tendency for the particles to be thrown out to the wall instead of tumbling upon each other and mixing. One can envision
a point where the induced forces on the particles result in their
being "glued" to the wall and no mixing occurs. On the other hand,
it is clear that if the Froude number is very low, very little tumbling of the particles occurs, and as a consequence the rate of mixing is very slow. Obviously an optimum rate of mixing exists at a
Froude number somewhere between the extremes of zero and a value
that causes the particles to be "glued" on the walls. This phenomenon is discussed in the literature (57). Observation of solids mixing reveals that there are three apparent regimes of solids movement
in rotation. At low Froude number, a "slug" flow exists where very
little mixing occurs because the material tends to move by breaking
large masses off the bulk. At intermediate Froude numbers, a gentle
rolling of the material occurs, and mixing of the particles is very
rapid. At very high Froude numbers, the material tends to "stick"
to the periphery of the mixer and very little mixing occurs.

Since there must be an optimum Froude number for mixing of the
particles, it can be concluded that the Froude number should be
kept constant at this optimum when scaling up a mixer. It falls on
pilot plant experimentation to determine what this optimum Froude
number is for the particular particles and mixer under consideration.
The conclusion that the Froude number should be fixed on scale-up
has been determined experimentally (57,58).

If the Froude number is fixed when scaling up a mixer, several
inferences about the design of the full-scale mixer can be made.
Since the Froude number is constant, the relationship between the
speed of the full-scale mixer and the pilot plant scale mixer is given
by

$$N \propto \frac{1}{D^{1/2}} \tag{152}$$

Thus, as the size of the mixer increases, the speed decreases. This
is consistent with the manufacturer's recommended speeds for tumbling blenders provided in Table 9.

If the Froude number is fixed on scale-up, then Eq. 151 implies
that the dimensionless blend time Nt is a constant on scale-up. It
can be shown that the relationship between mix time and scale of
the mixer is

$$t \propto D^{1/2} \tag{153}$$

The time required to perform the specified degree of mixing increases with vessel size. Unfortunately, this fact is often ignored
during process design.

TABLE 9   Recommended Speeds for Various Sizes of
Tumbling Blenders[a]

Capacity (cu ft)	Diameter (in.)/ turning radius (in.)[b]	Recommended speed (rpm)
Double cone blenders		
2	23	37
3	27	30
5	31	30
10	39	27
20	49	23
30	56	21.9
40	62	19.5
50	66	20
60	70	20
75	76	12
100	84	14.7
125	90	14
150	96	7
"V" shell blenders		
2	18.75	24.8
3	21.75	24.8
5	25	24
10	31.63	18.7
20	38.63	15.3
30	44.5	14.7
40	47.25	13.6
50	51.87	13.6
60	54.87	11.2
75	59.63	11
100	66.63	8.3
125	70	8.4
150	74	7

[a]This data is courtesy of the Patterson-Kelly Company,
East Stroudsburg, Pennsylvania.
[b]Diameter for double-cone blenders; turning radius for
"V" shell blenders.

This is a very clear example of the power of dimensional analysis as a tool for analyzing problems which defy conventional analysis. Through the use of dimensional analysis, and the judicious

application of intuition, a scale-up rule has been developed which would otherwise require much costly pilot plant experimentation.

Correctly, one might object to this analysis since in the general case neither the sizes or the densities of the particles being mixed will be uniform. In addition, there might be electrostatic or Van der Waals forces that affect the attraction of the particles to one another which will alter the mixing behavior of the system. These factors can be incorporated into the analysis.

For the general case of n components, each having their density size, and forces of attraction/repulsion between either particles of its own species or other species, the dimensional analysis proceeds as follows:

Variable	Definition	Dimensions
σ^2	Coefficient of variation	—
D	Mixer diameter	L
d_1	Diameter of particle 1	L
d_n	Diameter of particle n	L
ρ_1	Density of particle 1	M/L^3
.		
.		
ρ_n	Density of particle n	M/L^3
F_{11}	Attractive force between particle of type 1,1	M/LT^2
.		
.		
F_{nn}	Attractive force between particle of type n,n	M/LT^2
L	Mixer length	L
N	Mixer speed	1/T
t	Time of mixing	T
g	Gravitational acceleration	L/T^2
V	Volume of charge	L^3

Initially, this appears to be a very complicated problem. In fact it is not. There are $7 + 2n + n^2$ variables in the list and three fundamental dimensions, so $4 + 2n + n^2$ dimensionless variables are required to describe the process. The required equation is

$$\sigma^2 = f\left(\frac{N^2D}{g}, \frac{L}{D}, \frac{d_1}{D}, \frac{F_{11}}{\rho d_1^3 g}, \cdots \frac{F_{nn}}{\rho d_1^3 g}, \frac{\rho_1}{\rho_1} \cdots \frac{\rho_n}{\rho_1}, \right.$$

$$\left. \frac{d_2}{d_1} \cdots \frac{d_n}{d_1}, Nt, \frac{V}{D^3}\right) \tag{154}$$

The new terms that incorporate the attractive forces are similar to
a Froude number. Their physical interpretation is the ratio of at-
tractive/repulsive forces to gravitational forces.

Equation 154 appears to be formidable. If one were searching
for a general correlation, the task would be very difficult. From
a practical view, one does not need to go through this effort, since
for a particular design, there is only interest in one particular mixture of particles. As a result, all terms in Eq. 154 which incorporate attractive forces and the ratios of particle sizes or densities are
constants for the problem under consideration. This reduces Eq.
154 to

$$\sigma^2 = f\left(\frac{N^2D}{g}, \frac{V}{D^3}, \frac{L}{D}, Nt, \frac{d}{D}\right) \tag{155}$$

which is identical to Eq. 148. Once again, if geometric similarity
is maintained on scale-up and the particles are very small, the equation reduces to

$$\sigma^2 = f\left(\frac{N^2D}{g}, Nt\right) \tag{156}$$

which is identical to the result for the highly idealized case. The
conclusions drawn from the idealized case still apply to the very
general case.

Thus far, only a simple rotating drum has been considered. Although this type of device is used for the mixing of solids, the devices more commonly encountered will be considerably more complicated. For example, two very common types of solids mixing equipment used in the food industry are the ribbon blender (Fig. 24),
and a paddle mixer such as is found above a pasta extruder (Fig.
25). The analysis of these devices does not introduce a great deal
of additional difficulty.

Consider the ribbon blender of Figure 24. A list of variables
needed to describe this process is:

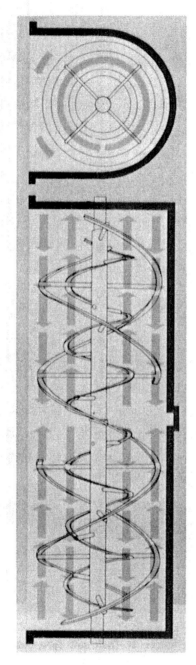

FIGURE 24 A ribbon blender. (Courtesy of J. H. Day Company, Cincinnati, Ohio.)

FIGURE 25   The paddle mixer at the feed to a pasta extruder.
(Courtesy of DeFranchisi Machinery Company.)

Variable	Definition	Dimensions
σ^2	Coefficient of variation	—
D	Mixer diameter	L
d	Diameter of particles	L
ρ	Density of particle	M/L^3
L	Mixer length	L
N	Mixer speed	$1/T$
t	Time of mixing	T
g	Gravitational acceleration	L/T^2
M	Mass of charge	M
T_o	Pitch of outer ribbon	L
D_o	Diameter of outer ribbon	L
W_o	Width of outer ribbon	L
T_i	Pitch of outer ribbon	L
D_i	Diameter of outer ribbon	L
W_i	Width of outer ribbon	L
z	Location of discharge	L
b	Ribbon/wall clearance	L

Comparison of this list with the list of variables for a simple drum reveals a similarity, with the addition of a considerable number of terms needed to describe the more complicated geometry of the ribbon blender. There are now a total of 17 variables and three fundamental dimensions, requiring the dimensional analysis to incorporate 14 dimensionless variables. The analysis is identical to that of the simple drum. All the new added dimensionless variables are simple ratios of length to a characteristic length (diameter). The result is

$$\sigma^2 = f\left(\frac{N^2 D}{g}, \; Nt, \; \frac{M}{\rho D^3}, \; \frac{d}{D}, \; \frac{L}{D}, \; \frac{T_o}{D}, \; \frac{T_i}{D}, \; \frac{D_o}{D}, \; \frac{D_i}{D}, \right.$$

$$\left. \frac{W_o}{D}, \; \frac{W_i}{D}, \; \frac{b}{D}, \; \frac{z}{D} \right) \tag{157}$$

Again, this appears to be a formidable equation to deal with. However, if the scale-up is restricted to geometrically similar mixers and small particles, the result reduces to

$$\sigma^2 = f\left(\frac{N^2D}{g}, Nt\right) \tag{158}$$

which is identical to the result for a simple drum. Of course, the analysis could have included the attractive/repulsive forces and all the appropriate particle sizes. The result would have been the same.

The simplification of Eq. 157 to Eq. 158 by restricting scale-up to geometrically similar cases is a strong argument for maintaining geometric similarity.

Obviously, if the analysis were performed on any other type of powder mixer, the results would have been the same as for Eq. 158.

Figure 26 illustrates some mixing data found in the literature (26), which have been plotted in the form suggested by Eq. 158. These data are for nearly ideal mixtures.

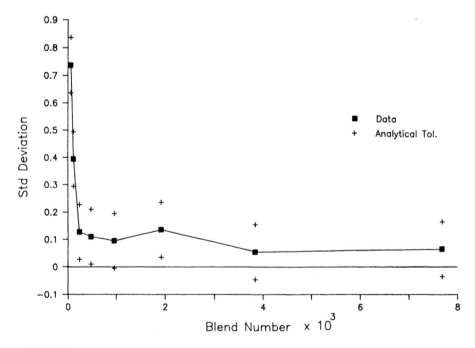

FIGURE 26   Mixing of a powder in a ribbon blender.   (From Ref. 26, used with permission.)

Figures 27 and 28 illustrate a commonly encountered experience. The two mixtures were very complicated formulas that incorporated many components having different densities and particle sizes. For these situations, it was discovered that if the product was mixed too long the materials would begin to segregate; as a result an optimal dimensionless mixing time of the key component was observed. In both figures, the optimal variance was plotted versus Froude number, as would be suggested by Eq. 154. An optimal Froude number exists for the process. In both cases, the optimum Froude number is approximately unity. Figure 29 is a plot of the dimensionless optimum mixing time versus the Froude number for the two systems. It is interesting to note that the slopes of these lines are very close to that suggested by Eqs. 152 or 153.

As with the tumbling drums, the relationship between speed and scale suggests that larger blenders should operate at a lower speed. This is suggested by a manufacturer's recommended ribbon blender speeds found in Table 10.

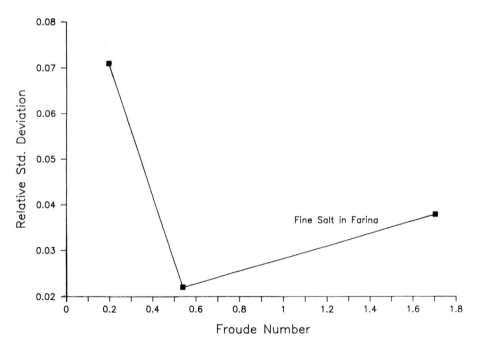

FIGURE 27 Mixing of fine salt in farina with a ribbon blender. (Courtesy of the Pillsbury Company.)

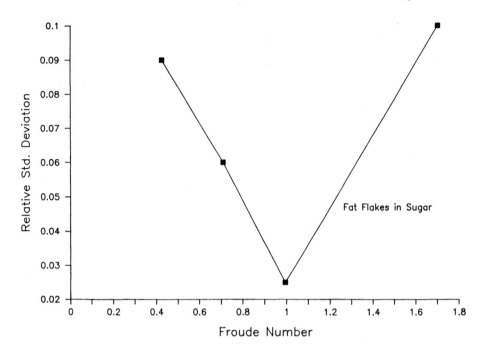

FIGURE 28 Mixing of fat flakes in sugar with a ribbon blender. (Courtesy of the Pillsbury Company.)

Before leaving the discussion of the mixing efficiency of in-powder mixers, some additional points should be made about mixing of multiple components.

If one or more components have a strong tendency to segregate, as was the case in the data presented above, it is often desirable to mix the bulk of the ingredients to some degree of uniformity and then to add the ingredients which have a tendency to segregate, near the end of the mixing cycle. The last ingredient will then be mixed just long enough so that it does not segregate. It is not difficult to see that on scale-up another factor must be maintained. For each minor ingredient the following addition rule must be maintained.

$$Nt_{addition} = constant \tag{159}$$

This helps explain a phenomenon often observed: The order of ingredient addition is important.

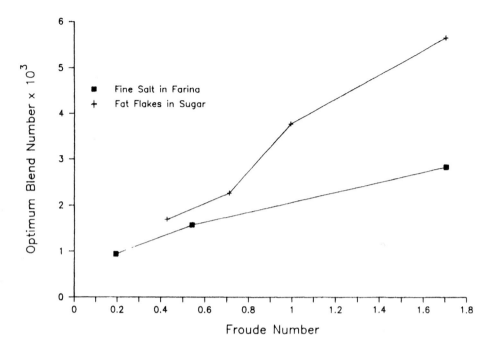

FIGURE 29   Correlation of optimum mix time with Froude number.

In reality, the addition of minor ingredients to the mixer may not occur instantaneously. A common practice is to charge the mixer with a major component, start the mixer, and then add the minor ingredients. In addition to the maintenance of the dimensionless time of addition described by Eq. 159, two additional variables are incorporated into the dimensional analysis: The rate of addition and the area over which the ingredient is distributed. This will add two additional groups, one describing a dimensionless addition rate (delivery number) and another describing dimensionless addition area. Experience has shown that these two factors and the time of addition factor are invariably ignored on scale-up. A complete dimensional analysis incorporating the addition of minor ingredients is

$$\sigma^2 = f\left(\frac{N^2 D}{g}, \frac{q_{add_2}}{ND^3} \cdots \frac{q_{add_n}}{ND^3}, Nt_{add_2} \cdots Nt_{add_n}, \right.$$

$$\left. \frac{A_2}{D} \cdots \frac{A_n}{D}, Nt + \text{geometric}\right) \tag{160}$$

TABLE 10   Recommended Speeds for Various Sizes of Ribbon
Blenders[a]

Blender capacity (cu ft)	Diameter (in.)	Length (in.)	Recommended speed (rpm)
5	17	38	60
10.8	21	48	60
17.5	24	60	60
23	26	66	50
36	30	78	50
54	34	90	45
62.5	36	96	40
80	40	96	25 (20)[b]
100	40	120	25 (20)
120	44	120	25 (20)
155	50	120	25 (20)
180	54	120	25 (20)
215	54	144	25 (20)
270	60	144	20 (15)
325	66	144	20 (15)
385	72	144	20 (15)
515	72	192	20 (15)

[a]This data is courtesy of the Day Mixing Company, Cincinnati, Ohio
(bulletin #DM-67 3M 11/86).
[b]Values in parentheses are for "heavy-duty" blenders.

If a specified degree of mixing is defined as for the case of idealized
mixing, the scale-up rule now becomes

$$Nt = f\left(\frac{N^2 D}{g}, \frac{q_{add_2}}{ND^3} \cdots, \frac{Nt_{add_2}}{ND^3} \cdots, \frac{A_2}{D} \cdots\right)$$ (161)

If the functional relationship suggested by Eq. 161 is not explicitly
known, successful scale-up can be accomplished by maintaining the
Froude number and the dimensionless additional groups that describe
the feeding of minor ingredients as constants on scale-up.

   Often the minor ingredients are liquids, added to control dust,
reduce segregation, or because the product formulation requires a
minor liquid ingredient. The analyses above are still valid provided
the quantity of liquid added is not so great as to convert the mass

into a liquid. Ignoring these additional factors related to feeding of minors is probably the most common cause of problems encountered when metering liquids into the system. Often the liquids are added with a simple pipe, or spray nozzle, which discharges onto the surface of the mixer. Equation 161 clearly shows that this is incorrect. The dimensionless term incorporating addition area suggests that as the process is scaled-up more pipe outlets or spray heads are required.

Besides simple mixing, all the arguments above apply to another unit operation often encountered in the food and pharmaceutical industries: the coating of one component with another by spraying or pouring the coating into a rotating drum. A common example of this is the pan coater schematically illustrated in Figure 30. There is one additional variable to consider in the process: the angle of the drum. This is a dimensionless variable that must be kept constant on scale-up.

K. Powder Mixer Power Consumption

In order to design processes that incorporate the tumbling of solids, the power consumption of these mixers must be predictable.

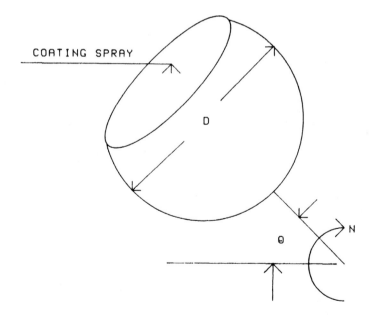

COATING SPRAY

FIGURE 30 A schematic representation of a pan coater.

Dimensional analysis combined with intuition provides insight into how power consumption may be predicted. The appropriate list of variables is:

Variable	Definition	Dimensions
P	Power consumption	ML/T^3
D	Mixer diameter	L
ρ	Density of particle	M/L^3
μ	Coefficient of friction	—
N	Rotational speed	$1/T$
g	Gravitational acceleration	L/T^2
M	Mass of charge	M

Since by now, the reader should be able to deal with the geometric terms that appear in these analyses, and should recognize that they are extraneous, the preceding list has omitted them.

The list includes the coefficient of friction for flow of the solids. This might be replaced by an angle of repose and/or other factors which describe the resistance of the solids to flow.

Since the list contains seven variable and three fundamental dimensions, three dimensionless groups are required to describe the system. By inspection, the result is

$$\frac{P}{\rho N^3 D^5} = f\left(\frac{N^2 D}{g}, \ \mu, \ \frac{m}{\rho D^3}\right) \tag{162}$$

If the degree of fill of the device is held constant, the result would be

$$\frac{P}{\rho N^3 D^5} = f\left(\frac{N^2 D}{g}, \ \mu\right) \tag{163}$$

Again, if only one material was of interest, the coefficient, which is a property of the material, is a constant and can be ignored.

It would be relatively easy to determine the constants of Eqs. 162 or 163 experimentally. This relationship would then be correct for any scale of mixer. Before exploring this option, consider where the intuition might lead.

Power dissipation is the result of frictional force that develops as objects slide by each other. For any two objects, such as powder mass moving relative to the wall of the mixer, the power consumed is the frictional force times the velocity. This infers that the power consumption of the mixer is proportional to the speed of rotation. Close examination of Eq. 163 reveals that this can only occur if the power number is inversely proportional to the Froude number, or

$$\frac{P}{\rho N^3 D^5} \propto \frac{g}{N^2 D} \tag{164}$$

Assuming that the intuitive argument above is correct, if the scale-up is restricted to a constant degree of fill, the scale-up relationship for power is established by Eq. 164. In principle, only one measurement of mixer power consumption is required to predict power consumption for any mixer scale.

The intuitive analysis could be carried further. What determines the friction force between two surfaces? The answer to this is, of course, the product of coefficient of friction and the normal force. In this case it seems that the normal force should be proportional to the mass of solids above the interface. This suggests that Eq. 164 should be written as

$$\frac{P}{\rho N^3 D^5} \propto \frac{g}{N^2 D} \cdot \frac{m}{\rho D^3} \tag{165}$$

or

$$\frac{P}{\rho N^3 D^5} \propto \frac{g}{N^2 D} \cdot \frac{\text{volume of fill}}{D^3} \tag{166}$$

Figures 31 and 32 should convince the reader of the benefits of this analysis. Figure 31 (49) describes the power consumption of a pan dryer used in the sugar industry. The relationship between power consumption, mass of charge, and speed is almost identical to that predicted by Eq. 165. The data of Figure 32 illustrates the same results for the paddle mixer found at the feed of pasta extruders (Fig. 25). Here, the agreement with the "theory" developed above is extraordinary. In addition, there is one data point for a very large mixer. This illustrates the validity of performing experimentation on only one scale if a dimensional analysis can be performed.

Another process that is sometimes encountered that is related to the mixing and coating analyses above is the grinding of material in ball mills. The literature (34) shows that experimental data for these devices agree with the arguments given above.

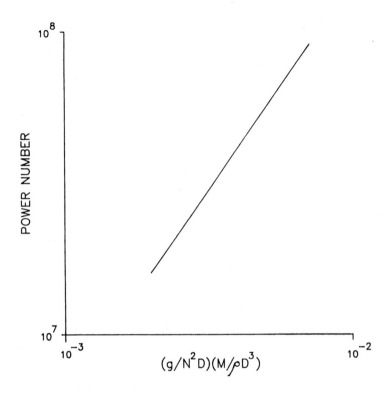

FIGURE 31   A correlation for power consumption for pan dryers.
(From Ref. 49, used with permission.)

V. CONCLUDING COMMENTS

We hope that Chapters 10 and 11 have provided the reader with
an appreciation of the problems associated with the scale-up of
food processes and approaches to solving these problems. Scale-
up of food processes is not an impossible task, but rather a series
of problems that are amenable to engineering analysis.

REFERENCES

1. Abichandani, H., S. C. Sarma, and D. R. Heldman, Hydrody-
    namics and heat transfer in liquid full scraped surface heat ex-
    changers—a review, *J. Food Proc. Eng.*, 9:121–141 (1987).

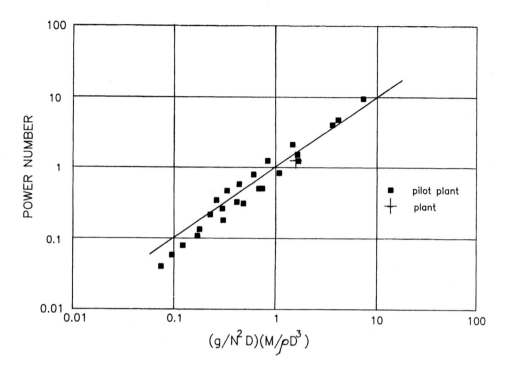

FIGURE 32 A correlation for power consumption by a twin shaft paddle mixer. (Courtesy of the Pillsbury Company.)

2. Bader, F. G., Improvements in multi-turbine mass transfer models, in *Biotechnology Progress*, C. S. Ho and J. Y. Oldshue, eds., American Institute of Chemical Engineers, New York (1987).
3. Bergen, J. T., and B. W. Scott, Pressure distribution in the calendering of plastic materials, *J. Appl. Mech.*, 18:101–106 (1951).
4. Bird, R. B., W. E. Stewart, and E. N. Lightfoot, *Transport Phenomenon*, John Wiley, New York (1960).
5. Bird, R. B., Armstrong, R. C., and O. Hassager, *Dynamics of Polymeric Fluids*, Vol. 1, John Wiley, New York (1977).
6. Bondy, F., and S. Lippa, Heat transfer in agitated vessels, *Chem. Eng.*, April:62–71 (1983).
7. Brazinsky, I., H. F. Cosway, C. F. Valle, R. Clark Jones, and V. Story, A theoretical study of liquid-film spread heights in the calendering of Newtonian and power law fluids, *J. Appl. Polym. Sci.*, 2:2771–2784 (1970).

8. Bridgeman, P. W., *Dimensional Analysis*, Yale University Press, New Haven (1922).
9. Carley, J. F., R. S. Mallouk, and J. M. McKelvey, Simplified flow theory for screw extruders, *Ind. Eng. Chem.*, 45(5):974–977 (1953).
10. Carley, J. F., and R. A. Strub, Application of theory to design of screw extruders, *Ind. Eng. Chem.*, 45(5):978–988 (1953).
11. Carslaw, H. S., and Jaeger, J. C., *Conduction of Heat in Solids*, Clarendon Press, Oxford (1986).
12. Cervon, N. W., and J. M. Harper, Viscosity of an intermediate moisture dough, *J. Food Process Eng.*, 2(2):83–95 (1978).
13. Charm, S. E., *Fundamentals of Food Engineering*, AVI, Westport, Connecticut (1981).
14. Chong, J. S., Calendering of thermoplastic materials, *J. Appl. Polym. Sci.*, 12:191–212 (1968).
15. Cuevas, R., M. Cheryan, and V. L. Porter, Performance of scraped surface heat exchanger under ultra high temperature condition: A dimensional analysis, *J. Food Sci.*, 47:619–625 (1982).
16. Della Valle, G., C. P. Kozlowski, and J. Tayed, Starch transformation estimated by the energy balance on a twin screw extruder, *Cereal Chemistry Manuscript*, in press.
17. Drew, B., L. Levine, V. Ramkrishna, and J. Clemmings, Comparison of mathematical models of dough sheeted through rolls, presented at the National Meeting of the American Institute of Chemical Enginners, August 17–19, Minneapolis, Minnesota (1987).
18. Drew, B., L. Levine, and V. Ramkrishna, Numerical solution to a problem in the flow of viscoelastic fluid between rotating cylinders, presented at the National Meeting of the Society for Industrial and Applied Mathematics, July 13–15, Minneapolis, Minnesota (1988).
19. Drew, B., and L. Levine, Rheological and engineering aspects of the sheeting and laminating of doughs, in *Dough Rheology and Baked Product Texture*, H. Faridi, ed., AVI Publishers, New York, in press.
20. Earle, R. L., *Unit Operation in Food Processing*, Pergamon Press, Oxford.
21. Ehrman, G., and J. Vlachopoulos, Determination of power consumption in calendering, *Rheol. Acta*, 14:761–764 (1975).
22. Elizarov, V. I., and T. K. Sirazetdinov, Non-Newtonian fluid flow in the gap between rotating cylinders. Isvestiya, *VUZ*, *Aviatsionnaya Teknika*, 16(4):10–16 (1973).
23. Faust, A. S., L. A. Wenzel, C. W. Clump, L. Maus, and L. B. Andersen, *Principles of Unit Operations*, John Wiley, New York (1960).

24. Finston, M., Thermal effects in the calendering of plastic materials, *J. Appl. Mech.*, 18(12):12–16 (1951).

25. Gaskell, R. E., The calendering of plastic materials, *J. Appl. Mech.*, 17:334–336 (   ).

26. Greathead, J. A. A., and W. H. C. Simmonds, Mixing patterns in helical-flight dry solids mixers, *Chem. Eng. Prog.*, 53(4): 194–198 (1957).

27. Harper, J. M., *Extrusion of Foods*, CRC Press, Boca Raton, Florida (1981).

28. Harper, J. M., Food extrusion, *CRC Critical Review in Food Science and Nutrition*, 11(2):155–215 (1979).

29. Heldman, D. R., and R. P. Singh, *Food Process Engineering*, AVI, Westport, Connecticut (1981).

30. Himmelblau, D. M., and K. B. Bischoff, *Process Analysis and Simulation, Deterministic Systems*, John Wiley, New York (1968).

31. Isaccson, E., and M. Issacson, *Dimensional Methods in Engineering and Physics*, John Wiley, New York (1975).

32. Janssen, L. P. B., *Twin Screw Extrusion*, Elsevier Science Publishing, New York (1978).

33. Jao, Y. C., A. H. Chen, D. Leandowski, and W. E. Irwin, Engineering analysis of soy dough, *J. Food Proc. Eng.*, 2(1): 97–112 (1978).

34. Johnston, R. E., and M. W. Thring, *Pilot Plant Models and Scale-up Methods in Chemical Engineering*, McGraw-Hill, New York (1957).

35. Kiparissides, C., and J. Vlachopoulos, Finite element analysis of calendering, *Polym. Eng. Sci.*, 16:712–719 (1976) .

36. Kline, S. J., *Similitude and Approximation Theory*, McGraw-Hill, New York (1965).

37. Levine, L., Throughput and power of dough sheeting rolls, *J. Food Proc. Eng.*, 7:223–228 (1985).

38. Levine, L., Estimating output and power of food extruders, *J. Food Proc. Eng.*, 6:1–13 (1982).

39. Levine, L., and J. Rockwood, A correlation of heat transfer coefficients in food extruders, *Biotechnol. Prog.*, 2(3);105–108 (1986).

40. Levine, L., and J. Rockwood, Simplified models for estimating isothermal operating characteristics of food extruders, *Biotechnol. Prog.*, 1(3):189 (1985).

41. Levine, L., S. Symes, and J. Weimer, A simulation of the effect of formula and feed rate variations on the transient behavior of starved extrusion screws, *Biotechnol. Prog.*, 3(4):221–230 (1987).

42. Marelli, F. G., *Twin Screw Extruders: A Basic Understanding*, Van Nostrand Reinhold, New York (1983).

43. McKelvey, J. M., *Polymer Processing*, John Wiley, New York (1962).

44. Middleman, S., *Fundamentals of Polymer Processing*, McGraw-Hill, New York (1981).
45. Mohamed, I. O., R. C. Morgan, and R. Y. Ofoli, Average convective heat transfer coefficients in single screw extruders of non-Newtonian food materials, *J. Food Proc. Eng.*, 2(1):68–75 (1986).
46. Mohesenin, N. M., *Thermal Properties of Food and Agricultural Materials*, Gordon and Breach, London (1980).
47. Morgan, R. G., D. A. Suter, and V. E. Sweat, Design and modeling of a capillary food extruder, *J. Food Proc. Eng.*, 2(1):65–81 (1978).
48. Oldshue, J. Y., *Fluid Mixing Technology*, McGraw-Hill, New York (1983).
49. *Perry's Chemical Engineering Handbook*, McGraw-Hill, New York (1963).
50. Remsen, C. H., and P. J. Clark, Viscosity model for a cooking dough, *J. Food Proc. Eng.*, 2(1):39–64 (1978).
51. Schenkel, G., *Plastics Extrusion Technology and Theory*, American Elsevier, New York (1966).
52. Stevens, M. J., *Extruder Principles and Operation*, Elsevier Applied Science, New York (1985).
53. Sears, F. W., and M. W. Zemansky, *University Physics, Part 2*, Addison-Wesley, Reading, Massachusetts (1964).
54. Tadmor, Z., and C. G. Gogos, *Principles of Polymer Processing*, John Wiley, New York (1979).
55. Tadmor, Z., and I. Klein, *Engineering Principles of Plasticating Extrusion*, Van Nostrand Reinhold, New York (1970).
56. Taylor, E. S., *Dimensional Analysis for Engineers*, Clarendon Press, Oxford (1974).
57. Uhl, V. W., and J. B. Gray, *Mixing Theory and Practice*, Academic Press, New York (1967).
58. Weidenbaum, S. S., Mixing of solids, in *Advances in Chemical Engineering*, Vol. II, ed. Drew and Hopkins, Academic Press, New York (1958).
59. Wilkinson, W. L., *Non-Newtonian Fluids*, Pergamon Press, New York (1960).
60. Zamodits, H. J., and J. R. A. Peason, Flow of polymer melts in extruders, *Trans. Soc. Rheol.*, 13(3):357–385 (1969).

Index